© 2024 by FAISAL JAMIL. All rights reserved.

Title: "The Impact of Climate Change on Global Food Production"

This book, along with its contents encompassing text, illustrations, images, diagrams, and other creative elements, is the exclusive property of FAISAL JAMIL and is safeguarded by copyright law.

FAISAL JAMIL asserts full ownership and retains all rights to this book. No part of this publication may be reproduced, distributed, or transmitted in any form or by any means, such as photocopying, recording, or electronic methods, without prior written consent from the copyright holder. Brief quotations in critical reviews and certain noncommercial uses permitted by copyright law are exceptions.

This copyright notice applies to all editions, formats, and translations of the book, whether in print, digital, or any other medium or technology existing now or developed in the future. Unauthorized use or infringement may result in legal action and pursuit of remedies under applicable copyright laws.

While efforts have been made to ensure accuracy and reliability, FAISAL JAMIL does not guarantee the completeness or suitability of the information. Readers are responsible for evaluating and using the content judiciously.

FAISAL JAMIL reserves the right to make changes, updates, or corrections to the book without prior notice. Inclusion of third-party materials or references does not imply endorsement or affiliation unless used under fair use principles or with proper permissions and attributions.

For permissions, inquiries, or requests regarding the book's use, please contact FAISAL JAMIL through official channels listed on their Amazon author page or provided email address.

This comprehensive copyright notice serves to protect FAISAL JAMIL'S intellectual property rights, maintain content control, and inform users about associated restrictions and permissions.

Warm regards,

FAISAL JAMIL

I Always Give's Free Copies Need Your Feedback And Reviews Keeps In Touch!

http://www.amazon.com/author/faisal.jamil

Email: faisaljamilauthor@gmail.com

About the author

Certainly! Faisal Jamil is a multifaceted individual with a diverse set of skills and experiences. With a strong foundation in computer knowledge since childhood, he has developed a deep understanding of technology that informs his work as a content writer. Faisal also possesses digital skills, which further enhance his abilities in various digital platforms and technologies.

Beyond his professional endeavors, Faisal Jamil has also excelled in the martial arts, particularly Shotokan Karate, where he achieved the prestigious rank of first Dan black belt. This achievement speaks to his dedication, discipline, and commitment to personal growth and mastery.

In his professional life, Faisal Jamil has carved out a successful career in sales management within the Fast Moving Consumer Goods (FMCG) sector. His roles in various FMCG companies have honed his skills in strategic planning, team leadership, and business development. Faisal's ability to drive sales and achieve targets has been instrumental in his career progression, showcasing his talent for identifying opportunities and delivering results.

Faisal Jamil is also deeply interested in business investment strategies, planning, and execution. His understanding of these areas has been key to his success in the business world, allowing him to make informed decisions and implement effective strategies. His ability to navigate the complexities of investment planning and execution has set him apart as a strategic thinker and a valuable asset in any business endeavor.

Overall, Faisal Jamil is a dynamic individual who combines his passion for technology, martial arts, sales management, digital skills, and business investment strategies to achieve success in diverse fields. His journey is a testament to his versatility, resilience, and continuous pursuit of excellence.

Yours Sincerely

FAISAL JAMIL

I Always Give's Free Copies Need Your Feedback And Reviews Keeps In Touch!

https://www.amazon.com/author/faisal.jamil

Email: faisaljamilauthor@gmail.com

THE IMPACT OF CLIMATE CHANGE
ON GLOBAL FOOD PRODUCTION

Table of Content

Preface	8
Introduction	10
Chapter 1: Introduction: The Climate-Food Nexus	13
Chapter 2: Historical Perspectives on Climate and Agriculture	17
Chapter 3: Current State of Global Food Production	22
Chapter 4: The Science of Climate Change	28
Chapter 5: Rising Temperatures and Crop Yields	34
Chapter 6: Extreme Weather Events and Agriculture	41
Chapter 7: Water Scarcity and Irrigation Challenges	48
Chapter 8: Soil Health and Climate Change	55
Chapter 9: Pests and Diseases in a Warming World	62
Chapter 10: The Impact on Livestock and Fisheries	69
Chapter 11: Regional Case Studies: Africa	78
Chapter 12: Regional Case Studies: Asia	85
Chapter 13: Regional Case Studies: Europe	92
Chapter 14: Regional Case Studies: Americas	99
Chapter 15: Technological Innovations in Agriculture	105
Chapter 16: Policy Responses to Climate Change	113

Chapter 17: Adaptation Strategies for Farmers   121

Chapter 18: The Role of Sustainable Practices   129

Chapter 19: Future Scenarios and Predictions   137

Chapter 20: Conclusion: Towards a Resilient

Food System   147

Preface

Embark on a compelling journey through "The Impact of Climate Change on Global Food Production," a vital exploration of how our planet's shifting climate is transforming agriculture and food security worldwide. As we face the unprecedented challenges of a warming world, understanding the delicate nexus between climate and food production is more critical than ever.

This book delves deep into the complexities of this relationship, tracing historical adaptations from ancient civilizations to the modern-day challenges confronting farmers across continents. By examining the science behind climate change—from the greenhouse effect to the role of human activities—we unravel its profound effects on crop yields, livestock, and fisheries.

With vivid regional case studies from Africa, Asia, Europe, and the Americas, this book brings to life the diverse impacts and innovative adaptation strategies employed around the globe. Learn about cutting-edge technological innovations like precision agriculture and genetically modified crops, alongside sustainable practices such as agroforestry and regenerative agriculture. Dive into policy responses and practical adaptation measures, highlighting the importance of coordinated global action and long-term planning.

Rich with insights and future predictions, "The Impact of Climate Change on Global Food Production" offers a hopeful yet realistic vision for a resilient food system. It is an essential read for anyone passionate about the future of our planet, the sustainability of our food sources, and the steps we must take to secure food security for generations to come.

INTRODUCTION

The relationship between climate and food production has always been a cornerstone of human civilization. From the fertile crescent of Mesopotamia to the rice terraces of Southeast Asia, our ability to cultivate and harvest food has been inextricably linked to the climatic conditions of our environment. Today, this delicate balance is under threat as climate change emerges as one of the most significant challenges of our time.

This book, "The Impact of Climate Change on Global Food Production," aims to explore the intricate nexus between our planet's shifting climate and the global food system. As the global population continues to grow, the demand for food rises, and the pressure on agricultural systems intensifies. Simultaneously, the effects of climate change—rising temperatures, altered precipitation patterns, increased frequency of extreme weather events, and shifting ecosystems—are profoundly altering the landscape of food production.

We begin by looking at historical perspectives, examining how ancient civilizations adapted their agricultural practices in response to changing climates. By learning from the past, we can gain insights into the resilience and innovation that will be necessary to address the challenges of today and tomorrow.

Next, we provide an overview of the current state of global food production, highlighting key statistics and trends. This foundation sets the stage for understanding how climate change is impacting the major crops and livestock that sustain the world's population, and which regions are most critical to global food security.

The scientific foundation of climate change is essential to our discussion, so we delve into the greenhouse effect, the carbon cycle, and the role of human activities in accelerating climate change. This knowledge underpins our exploration of the direct and indirect impacts of rising temperatures, extreme weather events, water scarcity, soil health, and the proliferation of pests and diseases on agriculture.

To illustrate the real-world impacts and adaptation strategies, we present detailed regional case studies from Africa, Asia, Europe, and the Americas. These case studies highlight the unique challenges faced by different regions and the innovative solutions being implemented to mitigate the effects of climate change on food production.

Technological innovations, such as precision agriculture and genetically modified crops, offer promising avenues for enhancing agricultural resilience. We examine these advancements alongside sustainable practices like agroforestry and regenerative agriculture, which play a crucial role in mitigating climate change impacts.

Policy responses at the international, national, and local levels are critical for addressing food security in the face of climate change. We explore how governments and institutions are crafting policies to support sustainable agricultural practices and promote resilience.

Adaptation strategies for farmers, including practical measures and case studies, provide a roadmap for building a resilient food system. The role of sustainable practices in mitigating the impacts of climate change is further emphasized, showcasing the potential for organic farming, agroforestry, and other regenerative methods to enhance food security.

Finally, we look to the future, presenting various scenarios and predictions for the impact of climate change on global food production. By examining potential outcomes based on current trends and possible mitigation and adaptation strategies, we aim to provide a comprehensive view of the challenges and opportunities that lie ahead.

"The Impact of Climate Change on Global Food Production" is not just an academic exploration; it is a call to action. Coordinated global efforts, innovative solutions, and long-term planning are essential to ensure food security for future generations. This book seeks to inspire and inform, offering a hopeful yet realistic vision for a resilient food system in the face of climate change.

Chapter 1
Introduction
The Climate-Food Nexus

The Historical Context

The relationship between climate and food production has been pivotal throughout human history. Ancient civilizations, such as those in Mesopotamia, Egypt, and the Indus Valley, flourished or declined based on their ability to adapt to their climatic conditions. For example, the success of Egyptian agriculture was closely tied to the annual flooding of the Nile River, which provided fertile soil for crops. Conversely, prolonged droughts contributed to the collapse of the Mayan civilization in Mesoamerica. These historical examples illustrate how climate has always played a critical role in shaping agricultural practices and food security.

Understanding the Climate-Food Relationship

The climate-food nexus refers to the interconnectedness between climate variables and food production systems. Key climate variables that influence food production include temperature, precipitation, atmospheric CO_2 levels, and the frequency and intensity of extreme weather events. Each of these factors can significantly impact crop growth, livestock health, and fishery productivity.

1: Temperature:

Optimal temperature ranges are crucial for the growth of specific crops. For instance, wheat and corn have distinct temperature requirements. Deviations from these optimal ranges can lead to reduced yields or crop failure.

2: Precipitation:

Adequate and timely rainfall is essential for crop irrigation. Both excess and deficient rainfall can harm crops, leading to issues such as waterlogging, drought stress, and soil erosion.

3: Atmospheric CO2 Levels:

While higher CO2 levels can enhance photosynthesis and potentially increase crop yields (a phenomenon known as the CO2 fertilization effect), this benefit can be offset by the adverse effects of higher temperatures and altered precipitation patterns.

4: Extreme Weather Events:

Climate change is increasing the frequency and severity of extreme weather events, such as hurricanes, floods, and droughts. These events can devastate agricultural infrastructure, reduce arable land, and disrupt food supply chains.

The Current State of Global Food Systems

Today's global food system is a complex network involving the production, processing, distribution, and consumption of food. It is influenced by various factors, including technological advancements, market dynamics, political policies, and socio-economic conditions. However, climate change is emerging as a dominant factor that threatens the stability and resilience of this system.

Global food production has increased significantly over the past few decades, driven by technological innovations and the expansion of agricultural land. However, this progress has come at a cost. Intensive farming practices have led to soil degradation, water scarcity, and loss of biodiversity. Additionally, the global food system is highly interdependent, meaning that climate-related disruptions in one region can have ripple effects worldwide.

The Urgency of Addressing Climate Change

The Intergovernmental Panel on Climate Change (IPCC) has highlighted the urgent need to address climate change to avoid catastrophic impacts on global food security. Rising temperatures, shifting precipitation patterns, and more frequent extreme weather events are already affecting food production. Without significant mitigation and adaptation efforts, these impacts are expected to intensify.

Key Challenges and Opportunities

Understanding the climate-food nexus is crucial for developing effective strategies to ensure food security in a changing climate. Key challenges include:

1: Adapting Agricultural Practices:

Farmers need to adopt more resilient agricultural practices, such as crop diversification, improved irrigation techniques, and soil conservation methods.

2: Technological Innovations:

Advances in biotechnology, precision agriculture, and climate modeling can help farmers adapt to changing conditions and optimize resource use.

3: Policy and Governance:

Effective policies at local, national, and international levels are essential to support sustainable agricultural practices, promote research and innovation, and ensure equitable access to resources.

4: International Cooperation:

Climate change is a global challenge that requires coordinated international efforts. Sharing knowledge, technologies, and resources can help build a more resilient global food system.

Conclusion

The delicate balance between climate and food production underscores the importance of understanding and addressing the impacts of climate change on global food systems. As we move forward, it is imperative to develop and implement strategies that enhance the resilience of agriculture, support sustainable practices, and ensure food security for future generations. This book aims to provide a comprehensive exploration of these issues, offering insights into the challenges and opportunities that lie ahead.

Chapter 2
Historical Perspectives on Climate and Agriculture

The Rise and Fall of Ancient Civilizations

Throughout history, climate variations have played a crucial role in the development and decline of civilizations. Ancient societies often thrived or faltered based on their ability to adapt to climatic changes. This chapter explores key historical examples to illustrate how climate influenced agricultural practices and food production, providing valuable lessons for addressing current and future challenges.

Mesopotamia: The Cradle of Civilization

Mesopotamia, often referred to as the "Cradle of Civilization," was home to some of the earliest known agricultural societies, including the Sumerians, Akkadians, Babylonians, and Assyrians. The region's agriculture was heavily reliant on the Tigris and Euphrates rivers, which provided the necessary water for irrigation in an otherwise arid environment.

1: Irrigation Innovations:

Ancient Mesopotamians developed sophisticated irrigation systems to control the flow of river water, enabling them to cultivate crops such as wheat, barley, and dates. These innovations allowed for surplus food production, supporting urbanization and the development of complex societies.

2: Climate Challenges:

Periodic droughts and shifting river courses posed significant challenges. Historical records and archaeological evidence suggest that severe droughts around 2200 BCE contributed to the collapse of the Akkadian Empire. This event underscores the vulnerability of agricultural systems to climate fluctuations.

Ancient Egypt: The Gift of the Nile

Ancient Egypt's agricultural success was closely tied to the annual flooding of the Nile River, which deposited nutrient-rich silt onto the surrounding land, creating fertile soil for crops.

1: Flood Cycles:

The predictability of the Nile's flood cycles allowed Egyptians to develop a calendar-based agricultural system, timing their planting and harvesting activities to coincide with the flooding.

2: Climate Variability:

While the Nile's floods were generally reliable, periods of low flood levels due to climate variations led to famine and social unrest. For instance, during the First Intermediate Period (circa 2181-2055 BCE), a series of low floods contributed to political instability and economic decline.

The Indus Valley Civilization

The Indus Valley Civilization, one of the world's earliest urban cultures, flourished in the northwest regions of South Asia around 2500-1900 BCE. The civilization's agriculture depended on the seasonal monsoons and the waters of the Indus River.

1: Agricultural Techniques:

Indus Valley farmers cultivated wheat, barley, peas, and cotton, and practiced crop rotation to maintain soil fertility. They also built extensive irrigation and drainage systems to manage water resources.

2: Climate Change and Decline:

Archaeological evidence suggests that the decline of the Indus Valley Civilization around 1900 BCE was partly due to climatic shifts that altered monsoon patterns, leading to reduced water availability and agricultural productivity.

The Mayan Civilization

The Mayan Civilization, known for its advanced knowledge of astronomy, mathematics, and architecture, thrived in Mesoamerica from approximately 2000 BCE to 1500 CE. Agriculture was central to Mayan society, with maize (corn) being the staple crop.

1: Agricultural Practices:

The Maya developed various agricultural techniques, including slash-and-burn (swidden) farming, terracing, and raised fields, to adapt to diverse ecological zones.

2: Drought and Societal Collapse:

Severe droughts in the 9th and 10th centuries CE are believed to have played a significant role in the collapse of Classic Maya civilization. Dendrochronological (tree ring) data and sediment cores indicate prolonged dry periods that would have devastated maize crops, leading to food shortages, social unrest, and the eventual abandonment of major urban centers.

Lessons from History

Examining these historical case studies reveals several key lessons about the relationship between climate and agriculture:

1: Adaptation and Innovation:

Ancient civilizations demonstrated remarkable adaptability and innovation in response to climatic challenges. Modern societies can learn from their irrigation techniques, crop diversification strategies, and resource management practices.

2: Vulnerability to Climate Variability:

Historical evidence highlights the vulnerability of agricultural systems to climate variability and extreme weather events. This underscores the need for robust and flexible agricultural practices to cope with changing climate conditions.

3: Socio-Political Impacts:

Climatic changes that disrupt food production can have profound socio-political impacts, including famine, migration, and conflict. Understanding these dynamics is crucial for developing strategies to maintain social stability in the face of climate change.

4: Sustainability and Resilience:

The sustainability and resilience of agricultural systems are vital for long-term food security. Ancient practices such as crop rotation, soil conservation, and water management remain relevant today as we seek to build resilient food systems.

Conclusion

The historical perspectives on climate and agriculture provide valuable insights into how ancient civilizations adapted to and were affected by climatic variations. These lessons from the past are essential for understanding the challenges we face today and for developing strategies to ensure a resilient and sustainable global food system in the context of ongoing and future climate change. As we move forward, it is crucial to draw on both historical knowledge and modern innovations to address the complex interplay between climate and agriculture.

Chapter 3
Current State of Global Food Production

Introduction

The global food production system is a complex and dynamic network that sustains the world's population. This chapter provides an overview of the current state of global food production, highlighting key statistics and trends. We will examine the major crops and livestock that form the backbone of the global food supply, and the regions that are most critical to ensuring food security.

Key Statistics and Trends

1: Global Food Production:

As of 2024, the global agricultural sector produces approximately 9.4 billion metric tons of food annually, encompassing crops, livestock, and fisheries.

The Food and Agriculture Organization (FAO) estimates that the world's population, currently around 8 billion, will reach 9.7 billion by 2050, necessitating a significant increase in food production.

2: Major Crops:

Cereals: Rice, wheat, and maize (corn) are the staple foods for over half of the world's population. Combined, these cereals account for about 50% of the global calorie intake.

Roots and Tubers: Potatoes, cassava, and sweet potatoes are vital for food security in many regions, particularly in Africa, Asia, and Latin America.

Legumes: Soybeans, lentils, and chickpeas are crucial for their protein content and are significant in both human and animal diets.

Fruits and Vegetables: These provide essential vitamins and minerals, with global production exceeding 2 billion metric tons annually.

Oil Crops: Palm oil, soy oil, and canola oil are key for both food and industrial uses, with palm oil being the most widely produced vegetable oil.

3: Livestock Production:

Livestock contributes approximately 40% of the global agricultural output in value terms.

The most widely produced livestock products include beef, pork, poultry, and dairy. The global meat production stands at about 350 million metric tons annually.

Poultry is the fastest-growing livestock sector, with chicken being the most consumed meat globally due to its affordability and shorter production cycle.

4: Fisheries and Aquaculture:

Fisheries and aquaculture provide around 17% of the global population's animal protein intake.

Global fish production is about 180 million metric tons annually, with aquaculture contributing more than half of this total, indicating a shift from traditional fishing to fish farming.

Regional Contributions to Global Food Security

1: Asia:

Asia is the largest producer of rice, contributing over 90% of the world's total production. Countries like China, India, and Indonesia are major producers.

The region also leads in the production of fruits, vegetables, and aquaculture products.

2: North America:

The United States and Canada are significant producers of maize, wheat, and soybeans. The U.S. is the world's largest exporter of corn and soybeans.

North America is also a major producer of beef and pork.

3: Europe:

The European Union is a key player in wheat production and a significant exporter of agricultural products.

Europe has a diversified agricultural sector, producing cereals, fruits, vegetables, and dairy products.

4: Latin America:

Brazil and Argentina are major producers of soybeans, beef, and poultry. Brazil is also the leading producer of sugarcane and coffee.

The region has vast agricultural land and favorable climates for diverse crop production.

5: Africa:

Africa is a major producer of cassava, yams, and sorghum. Nigeria is the largest producer of cassava globally.

The continent faces challenges such as low agricultural productivity and high post-harvest losses, but there is significant potential for growth.

6: Oceania:

Australia and New Zealand are prominent exporters of wheat, beef, and dairy products.

Oceania has a relatively small population but plays a crucial role in global food trade.

Challenges in Global Food Production

1: Climate Change:

Changing weather patterns, increased frequency of extreme weather events, and shifting growing seasons are impacting crop yields and livestock production.

Water scarcity and changing precipitation patterns are critical issues in many agricultural regions.

2: Resource Constraints:

Limited availability of arable land and water resources, along with soil degradation, are significant constraints on agricultural expansion.

Sustainable management of natural resources is essential to maintain and increase food production.

3: Technological and Infrastructure Gaps:

Many developing regions lack access to modern agricultural technologies and infrastructure, limiting their productivity and resilience.

Investment in agricultural research, technology dissemination, and rural infrastructure is crucial for improving food security.

4: Economic and Political Factors:

Trade policies, market fluctuations, and political instability can disrupt food production and distribution.

Ensuring fair and stable trade policies and strengthening political stability are important for global food security.

Opportunities for Enhancing Food Production

1: Technological Innovations:

Advances in biotechnology, such as genetically modified crops, can increase yields and resilience to pests and diseases.

Precision agriculture and digital farming technologies can optimize resource use and improve productivity.

2: Sustainable Practices:

Adopting sustainable agricultural practices, such as conservation tillage, crop rotation, and organic farming, can enhance soil health and reduce environmental impacts.

Integrating agroforestry and permaculture can diversify production systems and improve resilience.

3: Policy and Governance:

Implementing supportive policies that promote agricultural research, innovation, and sustainable practices is critical.

Strengthening international cooperation and creating frameworks for climate-resilient agriculture can enhance global food security.

Conclusion

The current state of global food production is characterized by significant achievements and persistent challenges. Understanding the major crops and livestock that sustain the world's population and the regions critical to global food security is essential for addressing the complex issues facing agriculture today. By leveraging technological innovations, sustainable practices, and effective policies, we can build a more resilient and sustainable global food system capable of meeting the needs of a growing population in the face of climate change and other challenges.

Chapter 4
The Science of Climate Change

Introduction

Understanding the science of climate change is fundamental to grasping its impact on global food production. This chapter delves into the key scientific concepts underlying climate change, including the greenhouse effect, the carbon cycle, and the role of human activities in accelerating climate change. By providing this scientific foundation, we can better appreciate the subsequent chapters that explore the specific effects of climate change on agriculture.

The Greenhouse Effect

1: What is the Greenhouse Effect?:

The greenhouse effect is a natural process that warms the Earth's surface. When the Sun's energy reaches the Earth, some of it is reflected back to space and the rest is absorbed, warming the planet.

The Earth then emits this absorbed energy as infrared radiation. Greenhouse gases (GHGs) in the atmosphere, such as carbon dioxide (CO_2), methane (CH_4), and water vapor (H_2O), trap some of this infrared radiation and prevent it from escaping back into space, thus warming the planet further.

2: Key Greenhouse Gases:

Carbon Dioxide (CO2): Released from burning fossil fuels, deforestation, and various industrial processes. It is the most significant GHG due to its abundance and long atmospheric lifetime.

Methane (CH4): Emitted during the production and transport of coal, oil, and natural gas, as well as from livestock and other agricultural practices. Methane is more effective than CO2 at trapping heat but exists in lower concentrations.

Nitrous Oxide (N2O): Emitted from agricultural and industrial activities, as well as during combustion of fossil fuels and solid waste.

Fluorinated Gases: Synthetic gases used in a variety of industrial applications. Although they are present in much smaller quantities, they are potent GHGs with a high global warming potential.

3: Impacts of the Enhanced Greenhouse Effect:

Human activities have significantly increased the concentrations of GHGs in the atmosphere, enhancing the natural greenhouse effect and leading to global warming.

The enhanced greenhouse effect is responsible for the observed increase in global average temperatures, changes in precipitation patterns, and more frequent and intense extreme weather events.

The Carbon Cycle

1; Overview of the Carbon Cycle:

The carbon cycle is the process by which carbon is exchanged among the atmosphere, oceans, soil, plants, and animals. It is a critical component of Earth's climate system.

Carbon exists in various forms, including CO_2 in the atmosphere, organic carbon in living organisms, and carbonate minerals in rocks.

2: Key Components of the Carbon Cycle:

Photosynthesis: Plants absorb CO_2 from the atmosphere and convert it into organic matter using sunlight. This process removes CO_2 from the atmosphere and stores it in plant biomass.

Respiration: Plants, animals, and microorganisms release CO_2 back into the atmosphere through respiration, a process that converts organic carbon back into CO_2.

Decomposition: When plants and animals die, their organic matter decomposes, releasing CO_2 and methane into the atmosphere.

Ocean-Atmosphere Exchange: The oceans absorb a large amount of CO_2 from the atmosphere. This CO_2 can be stored in the ocean for long periods or released back into the atmosphere.

Fossil Fuels: Over millions of years, organic matter can be buried and converted into fossil fuels (coal, oil, and natural gas). When these fuels are burned, the stored carbon is released back into the atmosphere as CO_2.

3: Human Impact on the Carbon Cycle:

Human activities, particularly the burning of fossil fuels and deforestation, have significantly altered the natural carbon cycle.

The increased CO_2 emissions from these activities are a major driver of the enhanced greenhouse effect and global warming.

Human Activities and Climate Change

1: Burning of Fossil Fuels:

The combustion of coal, oil, and natural gas for energy and transportation is the largest source of CO_2 emissions.

Industrial processes, such as cement production, also contribute to CO_2 emissions.

2: Deforestation and Land Use Changes:

Clearing forests for agriculture or urban development reduces the number of trees available to absorb CO_2 through photosynthesis.

Land use changes, such as converting forests to cropland, release stored carbon into the atmosphere and reduce the capacity of ecosystems to sequester carbon.

3: Agriculture:

Agricultural activities contribute to GHG emissions through livestock production (methane from enteric fermentation), rice paddies (methane emissions), and the use of synthetic fertilizers (nitrous oxide emissions).

Practices like slash-and-burn agriculture and the draining of wetlands also release significant amounts of carbon.

4: Industrial Activities:

The production of chemicals, metals, and other goods often involves processes that emit GHGs.

The release of fluorinated gases from industrial applications, although in smaller quantities, has a significant impact due to their high global warming potential.

Evidence of Climate Change

1: Temperature Records:

Global average temperatures have increased by approximately 1.2°C since the late 19th century, with the past decade being the warmest on record.

Temperature records from meteorological stations, satellites, and ocean buoys provide robust evidence of this warming trend.

2: Melting Ice and Rising Sea Levels:

Arctic sea ice extent has decreased significantly, and glaciers around the world are retreating.

The Greenland and Antarctic ice sheets are losing mass, contributing to global sea level rise, which has accelerated in recent decades.

3: Changes in Weather Patterns:

There is an increase in the frequency and intensity of extreme weather events, such as hurricanes, heatwaves, droughts, and heavy rainfall.

Changes in precipitation patterns are leading to wetter conditions in some regions and drier conditions in others.

4: Biological and Ecological Impacts:

Shifts in the distribution of plant and animal species have been observed, with some species moving to higher altitudes or latitudes in response to warming temperatures.

Changes in the timing of seasonal events, such as flowering and migration, are disrupting ecosystems.

Conclusion

The science of climate change is complex and multifaceted, involving the interplay of natural processes and human activities. The enhanced greenhouse effect, driven by increased concentrations of GHGs, is the primary cause of global warming. Understanding the carbon cycle and the significant impact of human activities on this cycle is crucial for addressing climate change. The evidence of climate change is clear and compelling, underscoring the need for urgent action to mitigate its impacts and adapt to the changes already underway. This scientific foundation sets the stage for exploring how climate change affects global food production and the strategies needed to ensure food security in a warming world.

Chapter 5
Rising Temperatures and Crop Yields

Introduction

Rising global temperatures are one of the most significant impacts of climate change, with profound effects on crop yields. This chapter examines how increased temperatures influence crop physiology, shift growing seasons, and affect agricultural productivity, particularly in regions most vulnerable to temperature changes.

The Physiological Impacts of Heat Stress on Plants

1: Photosynthesis and Respiration:

Photosynthesis: High temperatures can reduce the efficiency of photosynthesis, the process by which plants convert sunlight into chemical energy. The enzymes involved in photosynthesis, such as Rubisco, have optimal temperature ranges; beyond these ranges, their efficiency declines.

Respiration: Elevated temperatures increase the rate of respiration in plants, a process that consumes energy. Higher respiration rates can reduce the net energy available for growth and yield.

2: Water Use and Evapotranspiration:

Increased Water Demand: Higher temperatures lead to increased evapotranspiration, the sum of evaporation and plant transpiration from the Earth's surface. This increases water demand for crops.

Water Stress: In regions where water is already scarce, higher temperatures exacerbate water stress, leading to reduced growth and lower yields.

3: Heat Stress and Plant Development:

Flowering and Pollination: High temperatures can interfere with flowering and pollination, critical stages in the reproductive cycle of many crops. For instance, maize and wheat are highly sensitive to heat during their flowering periods.

Fruit and Grain Development: Heat stress can cause incomplete grain filling and lower quality in cereals. In fruits and vegetables, it can lead to sunscald and reduce marketability.

4: Pest and Disease Pressure:

Increased Pests: Warmer temperatures can expand the range and life cycle of many agricultural pests. For example, the fall armyworm, a significant pest in maize, has spread to new regions due to rising temperatures.

Disease Incidence: Higher temperatures can also increase the incidence and spread of plant diseases. Fungal pathogens, such as rusts and blights, thrive in warmer conditions.

Shifting Growing Seasons

1: Changes in Planting and Harvesting Times:

Earlier Planting: In many temperate regions, rising temperatures have led to earlier planting dates. While this can sometimes extend the growing season, it can also disrupt traditional farming schedules.

Accelerated Growth: Higher temperatures can accelerate plant growth, leading to earlier maturity and harvest. However, this can also mean that crops have less time to accumulate biomass, potentially reducing yields.

2: Growing Degree Days (GDD):

Concept of GDD: Growing Degree Days (GDD) is a measure of heat accumulation used to predict plant development rates. Rising temperatures increase GDD, affecting the timing of various growth stages.

Implications for Crop Selection: Farmers may need to select different crop varieties that are better suited to the new temperature regimes and altered GDD profiles.

3: Phenological Shifts:

Phenology: The study of periodic plant and animal life cycle events. Climate change is causing shifts in phenological events, such as earlier blooming and leaf-out dates in spring.

Mismatch in Timing: These shifts can lead to mismatches between crops and their pollinators, reducing pollination success and yields.

Regions Most Vulnerable to Temperature Changes

1: Tropical and Subtropical Regions:

Heat Stress: These regions already experience high temperatures, and further increases can push conditions beyond the tolerance limits of many crops.

Key Crops: Tropical crops like rice, maize, and cassava are particularly vulnerable. For example, rice yields in South and Southeast Asia are expected to decline significantly under higher temperature scenarios.

2: Temperate Regions:

Shifting Zones: In temperate regions, warmer temperatures may initially benefit some crops by extending the growing season. However, continued warming can lead to heat stress during critical growth stages.

Examples: Wheat production in Europe and North America may face challenges as optimal growing regions shift northward, necessitating changes in farming practices and crop varieties.

3: High-Latitude Regions:

Potential Benefits: Some high-latitude regions, such as parts of Canada and Russia, may experience increased agricultural potential as growing seasons lengthen and previously unsuitable areas become viable for farming.

Challenges: However, these benefits may be offset by soil quality issues, permafrost thawing, and increased pest pressures.

4: Drylands and Semi-Arid Regions:

Water Scarcity: In regions like the Middle East and North Africa, rising temperatures exacerbate water scarcity, a critical limiting factor for agriculture.

Adaptation Needs: Crops in these areas require substantial adaptation strategies, including drought-resistant varieties and improved irrigation practices.

Adaptation Strategies to Rising Temperatures

1: Crop Breeding and Biotechnology:

Heat-Tolerant Varieties: Developing and deploying crop varieties that are more tolerant of heat stress is a key adaptation strategy. Biotechnology can accelerate this process by introducing heat-tolerance traits.

Genetic Engineering: Advances in genetic engineering, such as CRISPR, offer potential for creating crops that can withstand higher temperatures and other climate-related stresses.

2: Improved Agricultural Practices:

Irrigation Management: Efficient irrigation techniques, such as drip irrigation, can help mitigate the increased water demand caused by higher temperatures.

Soil Health: Practices that improve soil health, such as cover cropping and reduced tillage, can enhance water retention and reduce temperature stress on crops.

3: Agroforestry and Mixed Cropping Systems:

Shade and Microclimates: Integrating trees and shrubs into agricultural landscapes can provide shade and create microclimates that protect crops from extreme heat.

Biodiversity: Mixed cropping systems can increase resilience by spreading risk and improving overall ecosystem health.

4: Policy and Research Support:

Investment in Research: Increased funding for agricultural research is critical to developing new technologies and practices that can help farmers adapt to rising temperatures.

Supportive Policies: Governments can implement policies that encourage the adoption of adaptive practices, provide financial support for affected farmers, and promote sustainable agriculture.

Conclusion

Rising global temperatures present a significant challenge to global food production. Understanding the physiological impacts of heat stress on plants, the shifting growing seasons, and the regions most vulnerable to temperature changes is essential for developing effective adaptation strategies. By leveraging advances in crop breeding, improved agricultural practices, and supportive policies, we can mitigate the negative effects of rising temperatures and ensure food security in a warming world.

This chapter sets the stage for exploring other climate-related impacts on agriculture and the comprehensive strategies needed to address these challenges.

Chapter 6
Extreme Weather Events and Agriculture

Introduction

Extreme weather events, such as hurricanes, droughts, and floods, are becoming more frequent and severe due to climate change. These events pose significant threats to agriculture and food production worldwide. This chapter explores the direct and indirect impacts of extreme weather on agriculture, examining how these events disrupt food systems and discussing strategies to mitigate their effects.

Increasing Frequency and Severity of Extreme Weather Events

1: Climate Change and Weather Extremes:

Rising Temperatures: Global warming contributes to more intense and prolonged heatwaves, increasing the likelihood of droughts and wildfires.

Altered Precipitation Patterns: Changes in atmospheric circulation patterns lead to more intense rainfall in some regions, causing floods, while other areas experience prolonged dry spells.

2: Key Types of Extreme Weather Events:

Hurricanes and Typhoons: These intense tropical storms are characterized by high winds, heavy rainfall, and storm surges. Their frequency and intensity are influenced by rising sea surface temperatures.

Droughts: Extended periods of low precipitation that can severely impact soil moisture, water supplies, and crop growth.

Floods: Excessive rainfall or rapid snowmelt can lead to riverine and flash floods, inundating agricultural land and infrastructure.

Heatwaves: Extended periods of excessively high temperatures that can cause heat stress in crops and livestock.

Direct Impacts on Agriculture

1: Crop Damage and Loss:

Hurricanes and Typhoons: High winds and heavy rains can cause physical damage to crops, uproot plants, and lead to waterlogging and soil erosion. For instance, Hurricane Maria in 2017 devastated agricultural production in Puerto Rico.

Floods: Floodwaters can inundate fields, suffocating crops, leaching nutrients from the soil, and depositing debris and contaminants. In 2010, floods in Pakistan submerged about one-fifth of the country's land area, destroying millions of hectares of crops.

Droughts: Prolonged dry conditions can lead to crop wilting, reduced yields, and total crop failure. The 2011 East Africa drought severely affected food production, leading to widespread famine.

2: Livestock Impacts:

Heat Stress: High temperatures can reduce feed intake, growth rates, and reproduction rates in livestock. Heat stress can also increase mortality rates, particularly in poultry and swine.

Water Scarcity: Drought conditions can limit water availability for livestock, affecting their health and productivity.

Flooding: Floodwaters can displace livestock, contaminate water supplies, and spread diseases.

3: Soil Degradation:

Erosion: Heavy rains and floods can erode topsoil, reducing soil fertility and agricultural productivity.

Salinization: Flooding, especially in coastal areas, can lead to soil salinization, making land unsuitable for most crops.

Indirect Impacts on Agriculture

1: Disruption of Supply Chains:

Transportation: Extreme weather can damage infrastructure, such as roads, bridges, and ports, disrupting the transportation of agricultural inputs and outputs.

Market Access: Farmers may be unable to access markets to sell their produce, leading to income losses and food shortages in affected areas.

2: Economic Losses:

Crop Insurance: Increased frequency of extreme weather events can lead to higher insurance premiums and reduced availability of crop insurance, making it more challenging for farmers to manage risk.

Investment: Uncertainty due to extreme weather can deter investment in agriculture, affecting long-term productivity and development.

3: Food Security and Nutrition:

Food Availability: Crop losses and livestock deaths can reduce food availability, leading to higher prices and food shortages.

Nutrition: Reduced availability of diverse and nutritious foods can lead to malnutrition, particularly in vulnerable populations.

Case Studies

1: Hurricane Maria (2017):

Impact on Puerto Rico: Hurricane Maria caused widespread destruction to Puerto Rico's agriculture, with estimated losses of $780 million. Coffee, plantain, and banana crops were particularly hard-hit.

Recovery Efforts: Efforts to rebuild the agricultural sector included replanting crops, restoring infrastructure, and providing financial assistance to farmers.

2: East Africa Drought (2011):

Impact on Food Production: The drought led to the failure of multiple consecutive harvests, exacerbating food insecurity and leading to a humanitarian crisis affecting over 13 million people.

Adaptation Strategies: Initiatives to build resilience included promoting drought-tolerant crops, improving water management, and supporting pastoralists with livestock feed and veterinary services.

3: Pakistan Floods (2010):

Extent of Damage: The floods affected over 20 million people and submerged around 17 million acres of agricultural land, leading to massive crop losses and displacement of livestock.

Response and Adaptation: Recovery measures included the distribution of seeds and fertilizers, rehabilitation of irrigation systems, and implementation of flood-resistant infrastructure.

Mitigation and Adaptation Strategies

1: Improved Weather Forecasting and Early Warning Systems:

Advanced Forecasting: Investing in meteorological infrastructure and technology to provide accurate and timely weather forecasts can help farmers prepare for extreme weather events.

Early Warning Systems: Implementing community-based early warning systems can ensure that farmers receive alerts and can take protective measures to minimize damage.

2: Climate-Resilient Agricultural Practices:

Diversified Cropping Systems: Planting a variety of crops can reduce the risk of total crop failure and enhance resilience to extreme weather events.

Agroforestry: Integrating trees and shrubs into agricultural landscapes can protect crops from wind damage, reduce soil erosion, and improve water retention.

3: Infrastructure Improvements:

Flood Defenses: Building levees, dams, and drainage systems can help protect agricultural land from flooding.

Irrigation Systems: Developing efficient and resilient irrigation systems can help manage water resources during droughts and ensure consistent water supply for crops.

4: Policy and Institutional Support:

Risk Management Tools: Expanding access to crop insurance and financial services can help farmers manage the economic impacts of extreme weather events.

Research and Development: Investing in agricultural research to develop climate-resilient crop varieties and farming techniques is crucial for long-term adaptation.

Conclusion

Extreme weather events, exacerbated by climate change, pose significant challenges to global agriculture and food production. The direct impacts on crops, livestock, and soil, along with the indirect effects on supply chains, economies, and food security, underscore the need for comprehensive mitigation and adaptation strategies. By improving weather forecasting, adopting climate-resilient agricultural practices, enhancing infrastructure, and supporting farmers through policies and financial tools, we can build a more resilient agricultural system capable of withstanding the increasing frequency and severity of extreme weather events. This chapter highlights the critical importance of addressing these challenges to ensure sustainable food production in the face of a changing climate.

Chapter 7
Water Scarcity and Irrigation Challenges

Introduction

Water is a fundamental resource for agriculture, underpinning crop and livestock production. However, climate change is exacerbating water scarcity, posing significant challenges to global food production. This chapter delves into the critical issue of water scarcity, examining how climate change affects freshwater availability, the challenges of irrigation, and potential solutions for sustainable water management in agriculture.

Water Scarcity and Climate Change

1: Definition and Causes of Water Scarcity:

Water Scarcity: A situation where water demand exceeds the available supply or when water quality restricts its use. It can be classified as physical (absolute) water scarcity or economic water scarcity.

Drivers of Water Scarcity: Population growth, urbanization, industrialization, and increased agricultural demands are major contributors. Climate change exacerbates these pressures by altering precipitation patterns and increasing evaporation rates.

2: Impact of Climate Change on Freshwater Availability:

Altered Precipitation Patterns: Climate change is causing shifts in precipitation, with some regions experiencing more intense rainfall and others facing prolonged droughts.

Melting Glaciers and Snowpacks: Glaciers and snowpacks, critical sources of freshwater for many regions, are melting at accelerated rates, affecting seasonal water availability.

Increased Evaporation: Rising temperatures increase evaporation from soil and water bodies, reducing the amount of water available for agriculture.

Changes in River Flows: Altered precipitation and melting patterns affect river flow regimes, impacting water availability downstream.

3: Regions Most Affected by Water Scarcity:

Arid and Semi-Arid Regions: Areas like the Middle East, North Africa, and parts of India and Australia are particularly vulnerable due to already limited water resources.

High-Demand Agricultural Areas: Regions with intensive agriculture, such as California's Central Valley and parts of China, face significant water scarcity challenges.

Small Island States: These regions are highly dependent on limited freshwater resources and are vulnerable to changes in precipitation and sea-level rise.

Irrigation Challenges

1: Current State of Irrigation:

Importance of Irrigation: Irrigation is crucial for enhancing agricultural productivity and ensuring food security, especially in arid and semi-arid regions.

Types of Irrigation Systems: Common methods include surface irrigation, sprinkler irrigation, and drip irrigation, each with its own advantages and challenges.

2: Challenges in Irrigation:

Water Use Efficiency: Many traditional irrigation systems are inefficient, with significant water losses due to evaporation, runoff, and seepage.

Salinization: Poor irrigation practices can lead to soil salinization, where excessive salts accumulate in the soil, reducing crop yields.

Infrastructure Limitations: Aging and poorly maintained irrigation infrastructure can lead to inefficiencies and water losses.

Access to Technology: Smallholder farmers often lack access to advanced irrigation technologies and practices, limiting their ability to manage water resources effectively.

3: Economic and Policy Constraints:

Cost of Irrigation Systems: High initial investment and maintenance costs can be prohibitive for many farmers, particularly in developing countries.

Water Pricing and Regulation: Ineffective water pricing and regulatory frameworks can lead to over-extraction and misallocation of water resources.

Competing Water Demands: Agriculture competes with domestic, industrial, and environmental needs, often leading to conflicts over water allocation.

Potential Solutions for Water Management in Agriculture

1: Improved Irrigation Technologies:

Drip Irrigation: A highly efficient system that delivers water directly to the plant roots, minimizing losses due to evaporation and runoff.

Sprinkler Systems: More efficient than surface irrigation, though less so than drip irrigation, and suitable for a variety of crops and terrains.

Smart Irrigation Systems: Utilize sensors and automated systems to optimize water use based on soil moisture, weather conditions, and crop needs.

2: Water-Saving Agricultural Practices:

Conservation Tillage: Practices that reduce soil disturbance, improve water retention, and reduce evaporation.

Mulching: Applying organic or synthetic mulch to soil surfaces can reduce evaporation, moderate soil temperature, and suppress weeds.

Crop Rotation and Diversification: Planting different crops with varying water needs and growth cycles can optimize water use and improve soil health.

3: Integrated Water Resource Management (IWRM):

Holistic Approach: IWRM promotes the coordinated development and management of water, land, and related resources to maximize economic and social welfare without compromising ecosystem sustainability.

Stakeholder Involvement: Engaging local communities, governments, and stakeholders in water management decisions ensures more equitable and effective outcomes.

4: Policy and Institutional Support:

Effective Water Pricing: Implementing water pricing mechanisms that reflect the true cost of water use and encourage conservation.

Incentives for Sustainable Practices: Providing financial incentives and technical support for farmers adopting water-saving technologies and practices.

Regulatory Frameworks: Strengthening regulations to prevent over-extraction and ensure sustainable water use.

5: Investment in Research and Development:

Crop Breeding: Developing drought-resistant crop varieties through conventional breeding and biotechnology.

Water Management Research: Investing in research to improve understanding of water resource dynamics and develop innovative management strategies.

Capacity Building: Training farmers and agricultural professionals in water management practices and technologies.

Case Studies

1: Israel's Water Management Success:

Technological Innovation: Israel has pioneered advanced irrigation technologies, such as drip irrigation, and desalination plants to address water scarcity.

Policy Framework: Strong regulatory policies, efficient water pricing, and incentives for water-saving technologies have contributed to Israel's success in managing water resources.

2: California's Central Valley:

Challenges: The Central Valley faces significant water scarcity due to over-reliance on groundwater, competition for water resources, and prolonged droughts.

Solutions: Investments in efficient irrigation systems, groundwater recharge projects, and water trading mechanisms have helped mitigate some of these challenges.

3: India's Water Management Initiatives:

Issues: India faces severe water scarcity in many regions, with inefficient irrigation practices and over-extraction of groundwater.

Programs: Initiatives such as the Pradhan Mantri Krishi Sinchayee Yojana (PMKSY) aim to enhance water use efficiency, expand irrigation coverage, and promote water conservation practices.

Conclusion

Water scarcity, driven by climate change and increasing demand, poses a significant threat to global agriculture. Addressing this challenge requires a multifaceted approach, including the adoption of efficient irrigation technologies, sustainable agricultural practices, integrated water resource management, and supportive policies and institutions. By investing in research, fostering innovation, and promoting equitable water management, we can enhance the resilience of agricultural systems and ensure food security in a changing climate. This chapter underscores the importance of proactive and collaborative efforts to manage water resources effectively and sustainably.

Chapter 8
Soil Health and Climate Change

Introduction

Soil health is a cornerstone of agricultural productivity and ecosystem sustainability. Climate change is exerting profound effects on soil health, influencing factors such as soil erosion, nutrient depletion, and the role of organic matter. This chapter explores how climate change impacts soil health and examines strategies to mitigate these effects to maintain and enhance soil fertility.

The Importance of Soil Health

1: Definition and Components of Soil Health:

Soil Health: The capacity of soil to function as a vital living ecosystem that sustains plants, animals, and humans. Healthy soil supports crop production, regulates water, and cycles nutrients.

Key Components: Soil structure, organic matter content, nutrient availability, water holding capacity, and biological activity.

2: Role in Agriculture:

Crop Growth: Healthy soil provides essential nutrients, water, and a stable structure for root development.

Ecosystem Services: Soil health is crucial for water filtration, carbon sequestration, and biodiversity.

Climate Change and Soil Health

1: Soil Erosion:

Increased Rainfall Intensity: Climate change is leading to more intense and erratic rainfall patterns, which can increase the risk of soil erosion. Heavy rains can wash away the topsoil, which is rich in organic matter and nutrients.

Wind Erosion: In arid and semi-arid regions, increased temperatures and reduced vegetation cover can enhance wind erosion, further degrading soil quality.

2: Nutrient Depletion:

Leaching: Increased rainfall can lead to greater leaching of essential nutrients like nitrogen, phosphorus, and potassium from the soil, reducing its fertility.

Volatilization: Higher temperatures can increase the volatilization of nitrogen from the soil, particularly in the form of ammonia, leading to nutrient loss.

3: Soil Organic Matter (SOM):

Decomposition Rates: Rising temperatures accelerate the decomposition of soil organic matter, reducing its content in the soil. This process releases carbon dioxide, contributing to greenhouse gas emissions.

Soil Structure: SOM is critical for maintaining soil structure, water retention, and nutrient availability. Its depletion can lead to poorer soil health and reduced crop productivity.

4: Soil Moisture and Water Holding Capacity:

Drought Conditions: Prolonged droughts reduce soil moisture, impairing plant growth and increasing the risk of soil erosion.

Water Retention: Healthy soil with high organic matter content has better water retention capacity, which is crucial for sustaining crops during dry periods. Climate change threatens this balance by disrupting precipitation patterns.

5: Salinization:

Sea-Level Rise: In coastal areas, rising sea levels can lead to the intrusion of saline water into freshwater aquifers and agricultural soils, causing salinization.

Irrigation Practices: Inefficient irrigation in arid regions can lead to the accumulation of salts in the soil, further degrading soil health.

Strategies to Mitigate Climate Change Impacts on Soil Health

1: Conservation Agriculture:

No-Till Farming: Reduces soil disturbance, helps maintain soil structure, and minimizes erosion. It also enhances water infiltration and retention.

Cover Cropping: Planting cover crops protects soil from erosion, adds organic matter, and improves nutrient cycling.

Crop Rotation: Alternating crops with different root structures and nutrient needs can improve soil health and reduce pest and disease pressures.

2: Soil Organic Matter Management:

Composting: Adding compost to soil increases organic matter content, improves soil structure, and enhances nutrient availability.

Green Manure: Growing and incorporating green manure crops into the soil adds organic matter and nutrients, improving soil health.

Biochar: Adding biochar (charcoal used as a soil amendment) can increase soil organic matter, improve water retention, and sequester carbon.

3: Erosion Control Measures:

Terracing: Creating terraces on slopes reduces runoff and soil erosion.

Buffer Strips: Planting grass or other vegetation strips along field edges and waterways can trap soil particles and reduce erosion.

Contour Farming: Plowing along the contour lines of a slope can reduce soil erosion and runoff.

4: Nutrient Management:

Integrated Nutrient Management: Combining organic and inorganic fertilizers to maintain soil fertility while minimizing nutrient losses.

Precision Agriculture: Using technology to apply fertilizers more efficiently and accurately based on soil nutrient levels and crop needs.

Cover Crops and Legumes: Incorporating legumes and other cover crops that fix nitrogen into crop rotations can naturally replenish soil nutrients.

5: Water Management:

Efficient Irrigation: Implementing water-efficient irrigation methods such as drip or sprinkler systems to reduce water use and prevent salinization.

Rainwater Harvesting: Capturing and storing rainwater for agricultural use can help mitigate the impacts of irregular rainfall patterns.

Soil Moisture Monitoring: Using sensors to monitor soil moisture levels can help optimize irrigation practices and reduce water stress on crops.

6: Salinity Management:

Leaching Excess Salts: Applying sufficient water to leach salts below the root zone can help manage soil salinity.

Salt-Tolerant Crops: Growing crops that are tolerant to higher salinity levels can sustain agricultural production in affected areas.

Improving Drainage: Enhancing soil drainage systems can prevent the accumulation of salts and improve soil health.

Case Studies

1: Conservation Agriculture in Brazil:

Adoption of No-Till Farming: Widespread adoption of no-till farming in Brazil has led to improved soil health, reduced erosion, and increased agricultural productivity.

Cover Cropping: Farmers in Brazil have successfully used cover crops to enhance soil organic matter and nutrient cycling.

2: Soil Organic Matter Management in the USA:

Composting and Organic Amendments: In regions like California, incorporating compost and other organic amendments into the soil has improved soil structure, water retention, and crop yields.

Biochar Applications: Research and field trials in the USA have demonstrated the benefits of biochar in enhancing soil health and sequestering carbon.

3: Erosion Control in the Philippines:

Terracing and Contour Farming: In the hilly regions of the Philippines, terracing and contour farming have significantly reduced soil erosion and improved agricultural sustainability.

Agroforestry Practices: Integrating trees and shrubs into agricultural landscapes has further protected soil from erosion and improved overall soil health.

Conclusion

Soil health is a critical component of sustainable agriculture and ecosystem resilience. Climate change poses significant threats to soil health through increased erosion, nutrient depletion, and loss of organic matter. However, by adopting conservation agriculture practices, improving soil organic matter management, implementing erosion control measures, optimizing nutrient and water management, and addressing salinity issues, we can mitigate these impacts and enhance soil fertility. These strategies are essential for ensuring the long-term sustainability of agricultural systems and food security in the face of a changing climate. This chapter underscores the importance of proactive soil management to adapt to and mitigate the effects of climate change on agriculture.

Chapter 9
Pests and Diseases in a Warming World

Introduction

As global temperatures rise and precipitation patterns shift due to climate change, the dynamics of agricultural pests and diseases are being profoundly affected. This chapter examines how these environmental changes influence the prevalence and distribution of pests and diseases, and explores the implications for crop protection and food security. Understanding these impacts is crucial for developing effective strategies to mitigate threats to agriculture.

The Influence of Climate Change on Pests and Diseases

1: Temperature and Pest Dynamics:

Increased Survival and Reproduction: Higher temperatures can accelerate the life cycles of many pests, leading to more frequent and larger outbreaks. Warmer winters may also reduce the mortality of pests that would normally be killed off by cold temperatures.

Expanded Ranges: Many pests are shifting their geographic ranges towards higher altitudes and latitudes in response to rising temperatures, exposing new areas to pest infestations. For instance, the mountain pine beetle has expanded its range northwards in North America.

Altered Timing: Changes in temperature can disrupt the synchronization between pests and their host plants, potentially leading to more severe infestations if pests emerge earlier in the growing season.

2: Precipitation Patterns and Disease Dynamics:

Increased Humidity and Disease Spread: Many plant diseases thrive in warm, humid conditions. Increased rainfall and humidity can promote the spread of fungal and bacterial diseases. For example, late blight in potatoes, caused by the fungus-like organism Phytophthora infestans, spreads more rapidly in wet conditions.

Drought Stress and Disease Susceptibility: Plants stressed by drought may become more susceptible to diseases. Reduced water availability can weaken plants, making them less able to resist infections.

3: Interactions Between Pests, Diseases, and Climate:

Complex Interactions: The interactions between pests, diseases, and climate are complex and can vary depending on the specific pest or pathogen, the host plant, and local environmental conditions. These interactions can lead to unpredictable outcomes, complicating efforts to manage pests and diseases.

Implications for Crop Protection

1: Increased Pest Pressure:

Higher Pest Populations: The increased survival and reproduction rates of pests can lead to higher pest populations, increasing the pressure on crops and requiring more intensive pest management efforts.

Resistance Development: More frequent use of pesticides to control higher pest populations can accelerate the development of pesticide resistance, making it more challenging to control pests effectively.

2: Emerging and Re-emerging Diseases:

New Disease Threats: Climate change can facilitate the emergence of new diseases or the re-emergence of previously controlled diseases, posing new challenges for crop protection. For instance, warmer temperatures have facilitated the spread of the wheat stem rust pathogen (Puccinia graminis f. sp. tritici) into new regions.

Increased Disease Incidence: Higher humidity and altered precipitation patterns can lead to more frequent and severe disease outbreaks, requiring new management strategies.

3: Impact on Crop Yields:

Yield Losses: Increased pest and disease pressure can lead to significant yield losses, threatening food security and farmers' livelihoods. For example, the fall armyworm (Spodoptera frugiperda), a highly destructive pest, has spread across Africa and Asia, causing substantial damage to maize and other crops.

4: Economic Costs:

Higher Management Costs: The need for more frequent and intensive pest and disease management can increase the economic burden on farmers. This includes the costs of pesticides, labor, and potential yield losses.

Market Impacts: Outbreaks of pests and diseases can disrupt agricultural markets, leading to price volatility and affecting food supply chains.

Strategies for Managing Pests and Diseases in a Changing Climate

1: Integrated Pest Management (IPM):

IPM Principles: IPM combines multiple strategies to manage pests and diseases sustainably. This includes biological control, cultural practices, resistant crop varieties, and the judicious use of pesticides.

Biological Control: Using natural predators and parasitoids to control pest populations can reduce reliance on chemical pesticides. For example, ladybugs are effective predators of aphids.

Cultural Practices: Crop rotation, intercropping, and maintaining healthy soil can reduce pest and disease pressure. Planting cover crops and practicing conservation tillage can also improve soil health and resilience.

2: Breeding for Resistance:

Resistant Varieties: Developing and planting crop varieties that are resistant to pests and diseases can provide long-term protection. Advances in biotechnology and genomics can accelerate the development of resistant crops.

Gene Editing: Techniques like CRISPR can be used to enhance crop resistance to pests and diseases by introducing specific genetic traits.

3: Climate-Smart Agriculture:

Adaptation Strategies: Implementing climate-smart agricultural practices can help farmers adapt to changing pest and disease dynamics. This includes diversifying crops, adjusting planting dates, and improving water management.

Early Warning Systems: Developing and utilizing early warning systems can help farmers anticipate and respond to pest and disease outbreaks. These systems rely on monitoring, data collection, and predictive modeling.

4: Policy and Extension Services:

Supportive Policies: Governments can support farmers by providing access to information, resources, and financial assistance for pest and disease management. Policies that promote sustainable agricultural practices and research are crucial.

Extension Services: Agricultural extension services play a key role in educating farmers about integrated pest management, resistant crop varieties, and climate-smart practices. Extension agents can also help disseminate early warning information and provide technical assistance.

Case Studies

1: Fall Armyworm in Africa:

Spread and Impact: The fall armyworm, native to the Americas, has spread rapidly across Africa, causing significant damage to maize and other staple crops. The pest's spread is facilitated by climate change and increased trade.

Management Strategies: Farmers and researchers are employing a combination of biological control agents, resistant crop varieties, and cultural practices to manage the pest. Early warning systems and regional cooperation are also critical.

2: Wheat Rust in the Middle East:

Emergence of New Strains: Climate change has contributed to the emergence of new, highly virulent strains of wheat rust in the Middle East. These strains threaten wheat production, a staple food in the region.

Breeding for Resistance: Collaborative efforts are underway to develop and deploy rust-resistant wheat varieties. International cooperation and research are essential to address this threat.

3: Coffee Leaf Rust in Central America:

Impact of Climate Change: Rising temperatures and changing precipitation patterns have increased the incidence of coffee leaf rust in Central America. This disease has devastated coffee plantations and affected the livelihoods of smallholder farmers.

Adaptation Measures: Farmers are adopting shade-grown coffee practices, improving crop management, and planting resistant coffee varieties to combat leaf rust. Support from governments and NGOs is crucial for these adaptation efforts.

Conclusion

Climate change is significantly altering the dynamics of agricultural pests and diseases, posing new challenges for crop protection and food security. The increased prevalence and distribution of pests and diseases require a comprehensive and adaptive approach to management. By embracing integrated pest management, breeding for resistance, climate-smart agriculture, and supportive policies and extension services, we can mitigate the impacts of pests and diseases in a warming world. This chapter underscores the importance of proactive and sustainable strategies to protect crops and ensure food security in the face of climate change.

Chapter 10
The Impact on Livestock and Fisheries

Introduction

Livestock and fisheries are vital components of the global food system, providing essential protein and livelihoods for billions of people worldwide. Climate change is posing significant challenges to these sectors, affecting animal health, productivity, and the sustainability of fish populations. This chapter delves into the impacts of climate change on livestock and fisheries, examining issues such as heat stress on animals, changes in fish populations, and the broader implications for protein supply and food security.

Climate Change and Livestock

1: Heat Stress on Livestock:

Physiological Effects: Higher temperatures can cause heat stress in livestock, leading to reduced feed intake, lower growth rates, decreased milk production, and impaired reproductive performance. For example, dairy cows exposed to heat stress produce less milk with lower fat and protein content.

Increased Mortality: Severe heat stress can increase mortality rates in livestock, particularly among poultry and pigs, which are more susceptible to high temperatures.

Behavioral Changes: Livestock may alter their behavior in response to heat, such as seeking shade, reducing activity levels, and changing grazing patterns, which can impact their overall health and productivity.

2: Water and Feed Availability:

Water Scarcity: Climate change can exacerbate water scarcity, affecting the availability of drinking water for livestock and water for forage and feed crop production. Reduced water availability can lead to dehydration, heat stress, and decreased productivity.

Feed Crop Production: Changes in precipitation patterns and increased temperatures can impact the growth and yield of feed crops, leading to feed shortages and increased feed costs. Droughts can reduce the availability of grazing land and forage, forcing farmers to rely on expensive supplemental feeds.

3: Disease and Parasite Prevalence:

Increased Disease Risk: Warmer temperatures and changing precipitation patterns can expand the range and prevalence of livestock diseases and parasites. For example, the spread of vector-borne diseases like bluetongue and Rift Valley fever is facilitated by warmer temperatures and increased rainfall.

Parasite Proliferation: Higher temperatures and humidity can enhance the reproduction and survival of parasites such as ticks, flies, and worms, increasing the burden of parasitic infections on livestock.

4: Genetic and Breeding Considerations:

Heat-Resilient Breeds: Developing and promoting heat-resilient livestock breeds that are better adapted to higher temperatures and changing climatic conditions can help mitigate the impacts of climate change.

Genetic Diversity: Maintaining genetic diversity within livestock populations is crucial for breeding programs aimed at improving resilience to climate change-related stresses.

Climate Change and Fisheries

1: Changes in Fish Populations:

Temperature Sensitivity: Fish are ectothermic animals, meaning their body temperature is regulated by the surrounding environment. Changes in water temperature can affect fish metabolism, growth rates, reproduction, and distribution. For example, many fish species are shifting their ranges poleward or to deeper waters in response to warming oceans.

Ocean Acidification: Increased CO_2 levels are causing ocean acidification, which can affect the growth and survival of marine organisms, particularly those with calcium carbonate shells and skeletons, such as mollusks and coral reefs. This can disrupt marine food webs and impact fish populations.

2: Habitat Degradation:

Coral Reefs and Mangroves: Coral reefs and mangroves are critical habitats for many fish species. Climate change-induced coral bleaching and mangrove degradation threaten these ecosystems, reducing biodiversity and fish populations.

Wetlands and Estuaries: Wetlands and estuaries, which serve as important breeding and nursery grounds for many fish species, are vulnerable to sea-level rise, increased salinity, and changing hydrological conditions.

3: Impact on Aquaculture:

Temperature Fluctuations: Aquaculture operations are sensitive to water temperature fluctuations, which can affect fish health, growth rates, and feed efficiency. Sudden temperature changes can lead to increased stress and mortality rates in farmed fish.

Disease and Parasite Outbreaks: Warmer water temperatures can promote the proliferation of pathogens and parasites in aquaculture systems, leading to disease outbreaks and significant economic losses.

Implications for Protein Supply and Food Security

1: Reduced Productivity:

Livestock Production: Heat stress, water scarcity, and feed shortages can reduce livestock productivity, leading to lower meat, milk, and egg yields. This can affect the availability and affordability of animal protein, particularly in regions heavily reliant on livestock for nutrition and income.

Fishery Yields: Changes in fish populations, habitat degradation, and overfishing can reduce fishery yields, impacting the availability of fish protein for human consumption. Declining fish stocks can also threaten the livelihoods of communities dependent on fishing.

2: Nutritional Impacts:

Protein Deficiency: Reduced availability of animal and fish protein can lead to nutritional deficiencies, particularly in vulnerable populations that rely on these sources for essential nutrients. Protein deficiency can have severe health implications, including impaired growth and development in children.

Micronutrient Availability: Livestock and fish are important sources of essential micronutrients such as iron, zinc, and omega-3 fatty acids. Reduced consumption of these foods can lead to deficiencies and associated health problems.

3: Economic and Social Impacts:

Livelihoods and Income: Livestock and fisheries are crucial for the livelihoods of millions of people worldwide. Climate change-induced disruptions to these sectors can lead to income losses, increased poverty, and social instability.

Market Volatility: Fluctuations in livestock and fishery yields can contribute to market volatility, affecting food prices and access. Price spikes in animal and fish products can disproportionately impact low-income consumers.

Strategies for Adaptation and Mitigation

1: Improved Animal Management:

Cooling Systems: Implementing cooling systems such as fans, misters, and shade structures can help mitigate heat stress in livestock.

Water Management: Ensuring adequate water supply and efficient water use practices can help address water scarcity challenges. Strategies include rainwater harvesting, improving water storage, and enhancing water distribution systems.

Nutrition and Feeding: Optimizing feeding strategies, including the use of high-quality feeds and supplements, can help improve livestock resilience to heat stress and feed shortages.

2: Sustainable Fisheries Management:

Protecting Marine Ecosystems: Conserving and restoring critical marine habitats such as coral reefs, mangroves, and seagrass beds can enhance the resilience of fish populations.

Adaptive Management: Implementing adaptive fisheries management practices, including monitoring and adjusting catch limits, can help sustain fish stocks in the face of changing environmental conditions.

Aquaculture Innovations: Developing resilient aquaculture systems, including the use of temperature-tolerant fish species, improved disease management, and sustainable feed sources, can enhance the sustainability of fish farming.

3: Policy and Institutional Support:

Research and Development: Investing in research to understand the impacts of climate change on livestock and fisheries and to develop innovative solutions is crucial. This includes breeding programs for resilient livestock and fish species, as well as improved management practices.

Extension Services: Strengthening agricultural and fisheries extension services can help disseminate knowledge and best practices to farmers and fishers, supporting their adaptation efforts.

Supportive Policies: Governments can implement policies that promote sustainable livestock and fisheries management, provide financial and technical support to farmers and fishers, and ensure food security through social protection programs.

Case Studies

1: Heat-Resilient Livestock Breeds in Africa:

Development and Adoption: In regions like sub-Saharan Africa, research institutions are developing heat-resilient livestock breeds such as the Red Maasai sheep and the Boran cattle, which are better adapted to high temperatures and water scarcity.

Impact: The adoption of these breeds has improved livestock productivity and resilience, contributing to food security and livelihoods in affected communities.

2: Sustainable Fisheries Management in the Philippines:

Community-Based Approaches: The Philippines has implemented community-based fisheries management programs that involve local fishers in decision-making processes, promote sustainable fishing practices, and protect marine ecosystems.

Outcomes: These initiatives have led to improved fish stocks, enhanced biodiversity, and increased community resilience to climate change impacts.

3: Aquaculture Adaptation in Bangladesh:

Innovative Practices: In response to rising temperatures and changing water conditions, Bangladeshi fish farmers are adopting innovative practices such as polyculture (raising multiple species in the same pond) and integrated rice-fish farming.

Benefits: These practices have increased fish production, diversified income sources, and improved food security in rural areas.

Conclusion

Climate change poses significant challenges to the livestock and fisheries sectors, impacting animal health, productivity, and fish populations. These changes threaten the availability of animal and fish protein, with profound implications for food security and livelihoods. By adopting adaptive strategies, promoting sustainable management practices, and supporting research and policy initiatives, we can mitigate the impacts of climate change on livestock and fisheries and ensure a stable and resilient protein supply for future generations. This chapter highlights the importance of proactive measures to protect these critical components of the global food system in a changing climate.

Chapter 11
Regional Case Studies: Africa

Introduction

Africa is one of the continents most vulnerable to the impacts of climate change, given its dependence on agriculture for livelihoods and food security. This chapter delves into how climate change is affecting food production across different regions of Africa. Through detailed case studies, we explore the unique challenges faced by various regions and the innovative adaptation strategies being employed to mitigate these impacts and ensure food security.

Climate Change and Agriculture in Africa

1: General Overview:

Agricultural Dependence: A significant portion of Africa's population relies on agriculture for their livelihoods, with smallholder farmers making up the majority. The sector is highly sensitive to climatic variations, making it particularly vulnerable to climate change.

Climate Sensitivity: Africa's diverse climates range from arid deserts to tropical rainforests, with each region facing distinct challenges due to climate change, such as droughts, floods, and shifts in growing seasons.

2: Major Climate Change Impacts:

Temperature Increases: Rising temperatures are causing heat stress in crops and livestock, reducing productivity and yield.

Precipitation Changes: Altered rainfall patterns lead to prolonged droughts in some regions and increased flooding in others, both of which adversely affect agricultural productivity.

Extreme Weather Events: The frequency and intensity of extreme weather events, such as cyclones and heavy rains, are increasing, causing damage to crops, livestock, and infrastructure.

Case Study 1: East Africa

1: Challenges:

Drought: Countries like Kenya, Ethiopia, and Somalia face recurrent droughts, which severely impact crop yields and livestock health.

Food Insecurity: Drought conditions lead to food shortages and increased reliance on food aid, exacerbating food insecurity.

2: Adaptation Strategies:

Drought-Resistant Crops: Farmers are adopting drought-resistant crop varieties such as sorghum, millet, and drought-tolerant maize to cope with water scarcity.

Irrigation Development: Initiatives like the construction of small-scale irrigation systems and rainwater harvesting techniques are helping farmers ensure water availability for crops during dry spells.

Agroforestry: Integrating trees into agricultural landscapes helps improve soil fertility, reduce erosion, and provide shade and moisture for crops.

3: Case Example:

Kenya's ASALs (Arid and Semi-Arid Lands): In Kenya's ASAL regions, farmers are increasingly turning to livestock that are better adapted to arid conditions, such as camels and goats, alongside growing drought-tolerant crops. Community-based natural resource management practices are also being promoted to sustain livelihoods.

Case Study 2: West Africa

1: Challenges:

Erratic Rainfall: Countries like Nigeria, Ghana, and Senegal experience unpredictable rainfall patterns, leading to either excessive flooding or prolonged dry periods.

Soil Degradation: Intensive farming practices, coupled with climate change, have led to soil degradation and reduced agricultural productivity.

2: Adaptation Strategies:

Conservation Agriculture: Practices such as minimum tillage, crop rotation, and cover cropping help preserve soil moisture and improve soil health.

Climate Information Services: Providing farmers with accurate and timely climate information helps them make informed decisions about planting and harvesting times, mitigating the impacts of erratic weather.

Integrated Pest Management: Addressing the increased prevalence of pests due to changing climates through integrated pest management (IPM) practices helps protect crop yields.

3: Case Example:

The Great Green Wall Initiative: This African-led movement aims to combat desertification and land degradation across the Sahel region by planting trees and restoring degraded lands. This initiative not only addresses environmental challenges but also supports agricultural productivity and food security.

Case Study 3: Southern Africa

1: Challenges:

Cyclones and Floods: Countries like Mozambique and Malawi face frequent cyclones and floods, causing widespread damage to crops and infrastructure.

Water Scarcity: The region also grapples with water scarcity, affecting both crop irrigation and livestock watering.

2: Adaptation Strategies:

Flood-Resilient Infrastructure: Building flood-resilient infrastructure, such as raised seedbeds and flood diversion channels, helps protect crops and reduce the impact of flooding.

Water Management: Implementing water-efficient irrigation techniques, such as drip irrigation, and improving water storage systems, such as constructing reservoirs and water pans, are crucial for managing water scarcity.

Diversification: Encouraging crop and income diversification helps farmers spread risk and maintain food security even in adverse climatic conditions.

3: Case Example:

Mozambique's Cyclone Adaptation: Following the devastating impact of Cyclone Idai, Mozambique has focused on building resilient agricultural practices, including the use of flood-tolerant crop varieties and constructing cyclone-proof storage facilities to protect food supplies.

Case Study 4: North Africa

1: Challenges:

Desertification: Countries such as Morocco, Algeria, and Egypt face significant challenges from desertification, which reduces the availability of arable land.

Water Scarcity: The region is characterized by low rainfall and high evaporation rates, making water scarcity a persistent issue.

2: Adaptation Strategies:

Efficient Water Use: Techniques such as drip and sprinkler irrigation, along with water recycling and reuse, are vital for conserving water in agriculture.

Sustainable Land Management: Practices such as terracing, contour farming, and the use of organic fertilizers help combat soil erosion and maintain soil fertility.

Crop Selection: Growing drought-tolerant and heat-resistant crops, such as barley, olives, and dates, helps farmers adapt to the harsh climatic conditions.

3: Case Example:

Morocco's Green Plan: Morocco's agricultural strategy, known as the "Green Plan," focuses on sustainable agricultural practices, efficient water use, and the development of high-value crops. The plan aims to increase agricultural productivity and resilience to climate change.

Case Study 5: Central Africa

1: Challenges:

Heavy Rains and Flooding: Countries like the Democratic Republic of Congo and Cameroon experience heavy rains that lead to flooding, soil erosion, and crop destruction.

Pests and Diseases: The region's humid climate creates favorable conditions for the proliferation of pests and diseases, impacting crop yields.

2: Adaptation Strategies:

Improved Drainage Systems: Constructing and maintaining effective drainage systems helps manage excess water and reduce the impact of flooding on agricultural lands.

Agroecological Practices: Implementing agroecological practices, such as intercropping, agroforestry, and the use of organic pest control methods, helps enhance resilience to pests and diseases.

Community-Based Approaches: Promoting community-based natural resource management and sustainable agricultural practices helps build local capacity to adapt to climate change.

3: Case Example:

Cameroon's Agroforestry Systems: In Cameroon, farmers are integrating trees and crops in agroforestry systems to improve soil fertility, reduce erosion, and provide additional sources of income. These systems enhance resilience to climate variability and contribute to food security.

Conclusion

Africa's agricultural sector is highly vulnerable to the impacts of climate change, with each region facing unique challenges. However, innovative adaptation strategies are being employed across the continent to mitigate these impacts and ensure food security. By adopting drought-resistant crops, improving water management, implementing sustainable land practices, and fostering community-based approaches, African farmers are building resilience to climate change. This chapter highlights the importance of these strategies and the need for continued support and investment to safeguard Africa's food production in the face of a changing climate.

Chapter 12
Regional Case Studies: Asia

Introduction

Asia, home to a vast and diverse range of agricultural systems, is experiencing significant impacts from climate change. From the rice paddies of Southeast Asia to the wheat fields of Central Asia, the continent faces a multitude of challenges that threaten food production and security. This chapter examines the impact of climate change on food production across different regions of Asia, highlighting specific challenges and adaptation strategies employed by various countries.

Climate Change and Agriculture in Asia

1: General Overview:

Agricultural Diversity: Asia's agriculture is incredibly diverse, ranging from intensive rice cultivation in the monsoon regions to extensive wheat farming in the drier interior. Each system has unique vulnerabilities to climate change.

Population Pressure: With a large and growing population, Asia faces immense pressure to maintain and increase food production amidst changing climatic conditions.

2: Major Climate Change Impacts:

Temperature Increases: Rising temperatures affect crop yields and livestock productivity, with heat stress posing a significant challenge.

Precipitation Changes: Changes in rainfall patterns lead to droughts in some areas and flooding in others, disrupting planting and harvesting cycles.

Extreme Weather Events: Increased frequency and intensity of extreme weather events, such as typhoons, cyclones, and heatwaves, cause direct damage to crops and infrastructure.

Case Study 1: South Asia

1: Challenges:

Monsoon Variability: Countries like India, Pakistan, and Bangladesh rely heavily on the monsoon season for water. Increasing variability and unpredictability of monsoon rains pose a severe threat to agricultural productivity.

Heatwaves: Rising temperatures and frequent heatwaves affect crop yields and livestock health, particularly in India and Pakistan.

2: Adaptation Strategies:

Climate-Resilient Varieties: Developing and adopting climate-resilient crop varieties, such as drought-resistant rice and heat-tolerant wheat, are crucial.

Water Management: Improving water management practices, such as rainwater harvesting, efficient irrigation systems, and groundwater recharge, helps mitigate the impacts of water scarcity.

Agroforestry: Integrating trees into agricultural systems helps improve soil health, provide shade, and enhance resilience to climatic variations.

3: Case Example:

India's National Initiative on Climate Resilient Agriculture (NICRA): NICRA aims to enhance the resilience of Indian agriculture to climate change through research and development, demonstration of climate-resilient technologies, and capacity building of farmers.

Case Study 2: Southeast Asia

1: Challenges:

Flooding and Sea-Level Rise: Countries like Vietnam, Thailand, and the Philippines face increased flooding and sea-level rise, threatening rice paddies and coastal agricultural areas.

Typhoons and Cyclones: The region is prone to frequent typhoons and cyclones, causing extensive damage to crops and infrastructure.

2: Adaptation Strategies:

Flood-Resistant Crops: Developing and promoting flood-resistant rice varieties help farmers maintain productivity during flood events.

Integrated Farming Systems: Combining crop, livestock, and aquaculture systems helps diversify income sources and spread risk.

Mangrove Restoration: Restoring mangroves along coastlines provides a natural barrier against storm surges and protects agricultural land from saltwater intrusion.

3: Case Example:

Vietnam's Climate-Smart Agriculture (CSA) Program: Vietnam is implementing CSA practices, including the use of stress-tolerant crop varieties, efficient irrigation techniques, and integrated pest management, to enhance agricultural resilience to climate change.

Case Study 3: East Asia

1: Challenges:

Urbanization and Land Use Changes: Rapid urbanization in countries like China and Japan leads to the loss of agricultural land and increased pressure on remaining farmland.

Water Scarcity: Northern China, in particular, faces severe water scarcity, affecting agricultural productivity.

2: Adaptation Strategies:

Water-Saving Technologies: Implementing water-saving technologies, such as drip irrigation and precision farming, helps optimize water use.

Urban Agriculture: Promoting urban agriculture and vertical farming in cities helps reduce the pressure on rural agricultural land and enhances food security.

Sustainable Land Management: Practices such as conservation tillage, crop rotation, and organic farming help maintain soil health and productivity.

3: Case Example:

China's South-to-North Water Diversion Project: This massive infrastructure project aims to transfer water from the water-rich south to the arid north, supporting agricultural production in water-scarce regions.

Case Study 4: Central Asia

1: Challenges:

Arid and Semi-Arid Conditions: Countries like Kazakhstan, Uzbekistan, and Turkmenistan face arid and semi-arid conditions, making agriculture highly dependent on irrigation.

Soil Salinization: Intensive irrigation practices lead to soil salinization, reducing soil fertility and crop yields.

2: Adaptation Strategies:

Efficient Irrigation: Implementing efficient irrigation techniques, such as drip and sprinkler systems, helps reduce water use and prevent soil salinization.

Salinity-Tolerant Crops: Developing and promoting salinity-tolerant crop varieties helps maintain productivity in saline soils.

Agroecological Practices: Implementing agroecological practices, such as crop diversification and agroforestry, enhances resilience to climatic variations.

3: Case Example:

Uzbekistan's Water Management Reforms: Uzbekistan is undertaking water management reforms, including improving irrigation infrastructure, adopting water-efficient technologies, and promoting community-based water management practices.

Case Study 5: The Himalayas

1: Challenges:

Glacial Melt and Water Availability: Countries like Nepal, Bhutan, and northern India rely on glacial meltwater for irrigation. Climate change-induced glacial melt threatens long-term water availability.

Landslides and Soil Erosion: Increased rainfall intensity leads to landslides and soil erosion, affecting agricultural land and infrastructure.

2: Adaptation Strategies:

Water Harvesting: Implementing water harvesting techniques, such as building small reservoirs and check dams, helps capture and store glacial meltwater.

Terrace Farming: Practicing terrace farming helps reduce soil erosion and landslide risks on steep slopes.

Agrobiodiversity: Promoting agrobiodiversity by cultivating a variety of crops enhances resilience to climatic variations and pest outbreaks.

3: Case Example:

Nepal's Community Forestry Program: Nepal's community forestry program involves local communities in managing forest resources, promoting sustainable land use, and enhancing resilience to climate change impacts.

Conclusion

Asia's diverse agricultural systems face significant challenges due to climate change, with each region experiencing unique impacts. However, innovative adaptation strategies are being employed across the continent to mitigate these impacts and ensure food security. From developing climate-resilient crop varieties and improving water management practices to promoting sustainable land use and community-based approaches, Asian farmers and policymakers are building resilience to climate change. This chapter underscores the importance of these strategies and the need for continued support and investment to safeguard Asia's food production in a changing climate.

Chapter 13
Regional Case Studies: Europe

Introduction

Europe, with its diverse climate zones and varied agricultural practices, is facing numerous challenges due to climate change. From the vineyards of Southern Europe to the dairy farms of Northern Europe, the continent's agriculture is undergoing significant transformations. This chapter provides a comprehensive look at how climate change is affecting agriculture across different regions of Europe, examining the impacts on both traditional and modern agricultural practices through detailed case studies.

Climate Change and Agriculture in Europe

1: General Overview:

Diverse Agricultural Systems: Europe's agricultural systems range from intensive, high-tech farms in Western Europe to traditional, small-scale farms in Eastern Europe.

Climate Zones: The continent spans multiple climate zones, from Mediterranean climates in the south to temperate and boreal climates in the north.

2: Major Climate Change Impacts:

Temperature Increases: Rising temperatures affect crop growth periods, yield quality, and livestock health.

Precipitation Changes: Shifts in precipitation patterns lead to droughts in some regions and excessive rainfall in others, impacting water availability and soil health.

Extreme Weather Events: Increased frequency of extreme weather events, such as heatwaves, storms, and floods, causes direct damage to crops and agricultural infrastructure.

Case Study 1: Southern Europe

1: Challenges:

Drought and Water Scarcity: Countries like Spain, Italy, and Greece are experiencing more frequent and severe droughts, leading to water scarcity.

Heatwaves: Rising temperatures and prolonged heatwaves stress crops and reduce yields, particularly in vineyards and olive groves.

2: Adaptation Strategies:

Drought-Resistant Crops: Developing and planting drought-resistant crop varieties, such as heat-tolerant grapes and olives, helps mitigate the impacts of water scarcity.

Efficient Irrigation: Implementing efficient irrigation techniques, such as drip irrigation and precision farming, optimizes water use and conserves resources.

Soil Management: Practices like mulching, conservation tillage, and organic amendments improve soil moisture retention and health.

3: Case Example:

Spain's Irrigation Innovations: In Spain, farmers are adopting advanced irrigation technologies, such as sensor-based systems and automated irrigation, to enhance water use efficiency in vineyards and orchards.

Case Study 2: Northern Europe

1: Challenges:

Increased Rainfall and Flooding: Countries like the UK, Ireland, and the Netherlands face increased rainfall and flooding, affecting crop and livestock production.

Changing Growing Seasons: Warmer temperatures are extending growing seasons, but also bringing new pests and diseases.

2: Adaptation Strategies:

Flood Management: Constructing flood defenses, such as dykes and drainage systems, helps protect agricultural land from flooding.

Crop Diversification: Diversifying crops and introducing new varieties that are resilient to changing conditions reduces the risk of crop failure.

Integrated Pest Management: Implementing integrated pest management (IPM) practices helps control new pests and diseases without relying heavily on chemical pesticides.

3: Case Example:

Netherlands' Water Management: The Netherlands is renowned for its innovative water management systems, including sophisticated drainage and flood control infrastructure, which protect its low-lying agricultural lands from flooding.

Case Study 3: Eastern Europe

1: Challenges:

Temperature Extremes: Countries like Poland, Ukraine, and Romania face temperature extremes, with hot summers and cold winters impacting crop productivity.

Soil Degradation: Intensive farming practices and climate change contribute to soil degradation and reduced fertility.

2: Adaptation Strategies:

Soil Conservation: Implementing soil conservation practices, such as cover cropping, crop rotation, and reduced tillage, helps maintain soil health and productivity.

Climate-Resilient Varieties: Developing and planting climate-resilient crop varieties, such as winter-hardy cereals and drought-tolerant vegetables, helps mitigate the impacts of temperature extremes.

Sustainable Farming Practices: Promoting sustainable farming practices, such as organic farming and agroecology, enhances resilience to climate change.

3: Case Example:

Poland's Agroecological Practices: In Poland, farmers are increasingly adopting agroecological practices, such as integrated crop-livestock systems and organic farming, to improve sustainability and resilience to climate change.

Case Study 4: Central Europe

1: Challenges:

Heatwaves and Drought: Countries like Germany, Austria, and Switzerland are experiencing more frequent heatwaves and droughts, affecting crop yields and livestock production.

Pest and Disease Pressure: Warmer temperatures are increasing pest and disease pressure on crops, particularly in viticulture and fruit orchards.

2: Adaptation Strategies:

Water Conservation: Implementing water conservation practices, such as rainwater harvesting and efficient irrigation, helps mitigate the impacts of drought.

Heat-Tolerant Varieties: Developing and planting heat-tolerant crop varieties, such as drought-resistant wheat and heat-tolerant maize, helps maintain productivity.

Integrated Pest Management: Using integrated pest management (IPM) techniques, including biological control and resistant varieties, helps manage increased pest and disease pressure.

3: Case Example:

Germany's Climate-Resilient Viticulture: German winemakers are adopting climate-resilient viticulture practices, such as planting heat-tolerant grape varieties and implementing precision irrigation, to cope with changing climatic conditions.

Case Study 5: Mediterranean Region

1: Challenges:

Desertification: Countries in the Mediterranean region, such as Spain, Italy, and Greece, face the threat of desertification due to prolonged droughts and land degradation.

Water Scarcity: The region is characterized by limited water resources and increasing competition for water between agriculture, industry, and urban areas.

2: Adaptation Strategies:

Sustainable Water Management: Implementing sustainable water management practices, such as rainwater harvesting, efficient irrigation systems, and wastewater reuse, helps mitigate water scarcity.

Agroforestry: Integrating trees into agricultural landscapes helps combat desertification, improve soil health, and enhance water retention.

Drought-Tolerant Crops: Developing and promoting drought-tolerant crop varieties, such as olives, almonds, and chickpeas, helps maintain productivity in arid conditions.

3: Case Example:

Italy's Olive Groves: Italian farmers are adopting drought-tolerant olive varieties and implementing sustainable water management practices to cope with increasing drought and water scarcity in the Mediterranean region.

Conclusion

Europe's agricultural sector is facing significant challenges due to climate change, with each region experiencing unique impacts. However, innovative adaptation strategies are being employed across the continent to mitigate these impacts and ensure food security. From developing climate-resilient crop varieties and improving water management practices to promoting sustainable land use and community-based approaches, European farmers and policymakers are building resilience to climate change. This chapter underscores the importance of these strategies and the need for continued support and investment to safeguard Europe's food production in a changing climate.

Chapter 14
Regional Case Studies: Americas

Introduction

The Americas, encompassing North, Central, and South America, present a wide array of climates and agricultural systems, each facing unique challenges due to climate change. This chapter explores the diverse impacts of climate change on food production across these regions, highlighting the specific challenges and innovative adaptation strategies being implemented to safeguard agricultural productivity and food security.

Climate Change and Agriculture in the Americas

1: General Overview:

Diverse Agricultural Systems: The Americas feature diverse agricultural systems, from large-scale industrial farming in North America to smallholder farms in Central and South America.

Climate Variability: The continent spans a wide range of climates, from temperate zones in North America to tropical and subtropical zones in Central and South America.

2: Major Climate Change Impacts:

Temperature Increases: Rising temperatures affect crop yields, livestock health, and the viability of certain crops in specific regions.

Precipitation Changes: Altered precipitation patterns lead to droughts in some areas and increased rainfall and flooding in others, impacting water availability and soil health.

Extreme Weather Events: Increased frequency and intensity of extreme weather events, such as hurricanes, droughts, and heatwaves, cause direct damage to agricultural infrastructure and crops.

Case Study 1: North America

1: Challenges:

Heatwaves and Drought: The United States and Canada are experiencing more frequent and severe heatwaves and droughts, affecting crop yields and water resources.

Changing Growing Seasons: Shifting growing seasons are impacting crop management and timing of planting and harvesting.

2: Adaptation Strategies:

Drought-Tolerant Crops: Developing and planting drought-tolerant crop varieties, such as drought-resistant corn and soybeans, helps mitigate the impacts of water scarcity.

Efficient Irrigation: Implementing advanced irrigation technologies, such as precision irrigation and sensor-based systems, optimizes water use and conserves resources.

Climate-Smart Agriculture: Adopting climate-smart agricultural practices, such as cover cropping, no-till farming, and diversified crop rotations, enhances resilience to climate change.

3: Case Example:

California's Climate Adaptation: California, a major agricultural hub, is implementing various adaptation strategies, including the development of drought-resistant crops, efficient irrigation systems, and policies to manage water resources sustainably.

Case Study 2: Central America

1: Challenges:

Hurricanes and Storms: Countries like Honduras, Guatemala, and Nicaragua face frequent hurricanes and storms, causing significant damage to crops and infrastructure.

Water Scarcity: Prolonged droughts and water scarcity pose major challenges for agriculture, particularly for smallholder farmers.

2: Adaptation Strategies:

Resilient Crops: Promoting the cultivation of resilient crops, such as drought-tolerant maize and beans, helps maintain productivity under adverse conditions.

Agroforestry: Integrating trees into agricultural systems helps improve soil health, provide shade, and protect against wind and water erosion.

Community-Based Approaches: Implementing community-based water management and disaster preparedness plans enhances local resilience to climate impacts.

3: Case Example:

Honduras' Agroforestry Initiatives: In Honduras, farmers are adopting agroforestry practices, combining coffee cultivation with tree planting to enhance soil health, reduce erosion, and improve water retention.

Case Study 3: South America

1: Challenges:

Amazon Deforestation: Brazil and other Amazonian countries face challenges related to deforestation, which exacerbates climate change impacts and threatens biodiversity and agricultural sustainability.

Temperature Extremes and Water Stress: Countries like Argentina and Chile face temperature extremes and water stress, affecting crop yields and livestock production.

2: Adaptation Strategies:

Sustainable Land Management: Implementing sustainable land management practices, such as conservation agriculture and integrated pest management, helps maintain soil health and productivity.

Water Management: Improving water management practices, such as rainwater harvesting and efficient irrigation systems, helps mitigate water scarcity.

Agrobiodiversity: Promoting agrobiodiversity by cultivating a variety of crops and traditional varieties enhances resilience to climate change and pest outbreaks.

3: Case Example:

Brazil's Sustainable Soybean Farming:

In Brazil, initiatives are underway to promote sustainable soybean farming practices, including zero-deforestation commitments, integrated pest management, and soil conservation techniques.

Case Study 4: Caribbean

1: Challenges:

Hurricanes and Sea-Level Rise:

Caribbean islands face increasing threats from hurricanes and sea-level rise, which damage agricultural infrastructure and reduce arable land.

Saltwater Intrusion:

Sea-level rise leads to saltwater intrusion into freshwater resources, affecting crop irrigation and soil health.

2: Adaptation Strategies:

Salt-Tolerant Crops:

Developing and promoting salt-tolerant crop varieties, such as salt-resistant rice and vegetables, helps maintain agricultural productivity.

Coastal Protection:

Implementing coastal protection measures, such as mangrove restoration and construction of seawalls, helps protect agricultural land from sea-level rise and storm surges.

Disaster Preparedness:

Strengthening disaster preparedness and response systems helps mitigate the impacts of hurricanes and extreme weather events on agriculture.

3: Case Example:

Barbados' Coastal Agriculture:

In Barbados, farmers are adopting salt-tolerant crop varieties and implementing coastal protection measures, such as mangrove planting, to safeguard agricultural land from sea-level rise and saltwater intrusion.

Conclusion

The Americas face diverse and significant challenges due to climate change, affecting agriculture across North, Central, and South America as well as the Caribbean. However, innovative adaptation strategies are being employed to mitigate these impacts and ensure food security. From developing climate-resilient crop varieties and improving water management practices to promoting sustainable land use and community-based approaches, farmers and policymakers across the continent are building resilience to climate change. This chapter underscores the importance of these strategies and the need for continued support and investment to safeguard food production in the Americas in a changing climate.

Chapter 15
Technological Innovations in Agriculture

Introduction

As climate change intensifies, the agricultural sector faces increasing challenges in maintaining and enhancing food production. Technological innovations offer critical solutions to these challenges, enabling farmers to adapt to changing conditions, optimize resource use, and increase resilience. This chapter delves into various technological advancements in agriculture, focusing on precision agriculture, genetically modified crops, and sustainable farming practices.

Precision Agriculture

1: Overview:

Definition:

Precision agriculture involves the use of advanced technologies to monitor and manage field variability in crops, optimizing inputs like water, fertilizers, and pesticides to enhance productivity and sustainability.

Benefits:

This approach improves efficiency, reduces waste, and increases crop yields while minimizing environmental impacts.

2: Technologies:

GPS and GIS:

Global Positioning Systems (GPS) and Geographic Information Systems (GIS) enable precise mapping and monitoring of fields, allowing farmers to manage spatial variability more effectively.

Remote Sensing:

Drones and satellites equipped with sensors provide real-time data on crop health, soil conditions, and weather patterns, helping farmers make informed decisions.

IoT and Smart Sensors:

Internet of Things (IoT) devices and smart sensors collect data on soil moisture, temperature, and nutrient levels, facilitating precision irrigation and fertilization.

Data Analytics and AI:

Advanced data analytics and artificial intelligence (AI) tools analyze large datasets to provide predictive insights and recommendations for optimal farming practices.

3: Case Example:

John Deere's Precision Farming Solutions:

John Deere offers a suite of precision farming technologies, including GPS-guided tractors, variable rate technology for planting and fertilizing, and advanced data analytics platforms that help farmers optimize their operations.

Genetically Modified Crops

1: Overview:

Definition:

Genetically modified (GM) crops are plants whose DNA has been altered using genetic engineering techniques to introduce desirable traits, such as pest resistance, herbicide tolerance, and improved nutritional content.

Benefits:

GM crops can enhance yield, reduce the need for chemical inputs, and improve resilience to environmental stresses, contributing to food security in the face of climate change.

2: Technologies:

Gene Editing:

Techniques like CRISPR-Cas9 allow precise editing of plant genomes to introduce or modify specific traits, such as drought tolerance or disease resistance.

Transgenic Crops:

Traditional genetic engineering methods involve the insertion of genes from other species to confer beneficial traits, such as Bt cotton and Bt corn, which produce their own insecticides.

3: Case Example:

Drought-Tolerant Maize in Africa:

The Water Efficient Maize for Africa (WEMA) project developed drought-tolerant maize varieties using both conventional breeding and genetic modification, helping farmers in drought-prone regions maintain productivity.

Sustainable Farming Practices

1: Overview:

Definition:

Sustainable farming practices aim to enhance agricultural productivity while minimizing environmental impacts and ensuring long-term sustainability of natural resources.

Benefits:

These practices promote soil health, conserve water, reduce greenhouse gas emissions, and enhance biodiversity, contributing to the resilience of farming systems.

2: Technologies:

Conservation Tillage:

Reducing or eliminating tillage helps preserve soil structure, reduce erosion, and enhance carbon sequestration.

Cover Cropping:

Planting cover crops during off-seasons improves soil health, prevents erosion, and enhances nutrient cycling.

Agroecology and Agroforestry:

Integrating trees and other perennial plants into farming systems promotes biodiversity, improves soil health, and enhances resilience to climatic changes.

Integrated Pest Management (IPM):

Combining biological, cultural, mechanical, and chemical methods to manage pests reduces reliance on synthetic pesticides and minimizes environmental impact.

3: Case Example:

Agroforestry in Kenya:

In Kenya, farmers are adopting agroforestry practices, such as planting trees alongside crops, which improves soil fertility, enhances water retention, and provides additional sources of income through timber and fruit production.

Innovative Water Management

1: Overview:

Importance:

Effective water management is crucial for agriculture, especially in the context of increasing water scarcity and variability due to climate change.

Benefits:

Innovative water management practices enhance water use efficiency, reduce waste, and ensure a reliable water supply for crops and livestock.

2: Technologies:

Drip Irrigation:

Delivering water directly to the root zone of plants minimizes evaporation and runoff, improving water use efficiency.

Rainwater Harvesting:

Collecting and storing rainwater for agricultural use helps supplement irrigation during dry periods.

Soil Moisture Sensors:

Sensors that monitor soil moisture levels in real-time enable precise irrigation scheduling, reducing water use while maintaining optimal soil conditions for crop growth.

3: Case Example:

Israel's Water-Saving Technologies:

Israel, a leader in water management, has developed and implemented various technologies, such as drip irrigation and wastewater recycling, to maximize water use efficiency in agriculture.

Digital Farming Platforms

1: Overview:

Definition:

Digital farming platforms integrate various technologies and data sources to provide comprehensive farm management solutions, enhancing decision-making and operational efficiency.

Benefits:

These platforms offer real-time insights, facilitate data-driven decisions, and streamline farm management processes, improving productivity and sustainability.

2: Technologies:

Farm Management Software:

Software solutions that aggregate and analyze data from various sources, such as weather forecasts, soil sensors, and crop monitoring systems, help farmers optimize their operations.

Mobile Apps:

Mobile applications provide farmers with easy access to information, alerts, and recommendations, enabling them to manage their farms more effectively.

Blockchain for Traceability:

Blockchain technology enhances transparency and traceability in the supply chain, ensuring food safety and authenticity.

3: Case Example:

Climate FieldView:

Climate FieldView, a digital farming platform by The Climate Corporation, offers comprehensive farm management tools, including data analytics, field monitoring, and variable rate application, helping farmers optimize inputs and improve yields.

Conclusion

Technological innovations are playing a pivotal role in addressing the challenges posed by climate change in agriculture. Precision agriculture, genetically modified crops, sustainable farming practices, innovative water management, and digital farming platforms are among the key advancements enabling farmers to adapt to changing conditions, optimize resource use, and increase resilience. These technologies not only enhance agricultural productivity and sustainability but also contribute to global food security in a rapidly changing climate. This chapter highlights the importance of continued research, development, and adoption of innovative agricultural technologies to ensure a resilient and sustainable future for agriculture.

Chapter 16
Policy Responses to Climate Change

Introduction

As climate change poses an increasing threat to global food security, policy responses at international, national, and local levels become crucial in mitigating its impacts. This chapter explores the array of policy measures and strategies adopted by governments and international organizations to address climate change, focusing on food security. We will examine the effectiveness of these policies and highlight best practices from different regions.

International Policy Responses

1: United Nations Framework Convention on Climate Change (UNFCCC):

Overview:

The UNFCCC, established in 1992, provides a framework for international cooperation on climate change. It aims to stabilize greenhouse gas concentrations to prevent dangerous anthropogenic interference with the climate system.

Key Agreements:

The Kyoto Protocol and the Paris Agreement are critical milestones under the UNFCCC. The Paris Agreement, in particular, sets targets for limiting global temperature rise and enhancing adaptive capacity.

2: Paris Agreement (2015):

Goals:

The agreement aims to keep global temperature rise well below 2°C above pre-industrial levels and to pursue efforts to limit the increase to 1.5°C.

Nationally Determined Contributions (NDCs):

Countries submit NDCs outlining their climate actions, including measures to enhance food security. These plans are revised every five years to increase ambition.

Global Stocktake:

The periodic assessment of collective progress towards achieving the goals of the Paris Agreement, guiding future action and policy adjustments.

3: Sustainable Development Goals (SDGs):

Goal 2: Zero Hunger:

This goal aims to end hunger, achieve food security and improved nutrition, and promote sustainable agriculture by 2030. It is closely linked to climate action (Goal 13) and sustainable water management (Goal 6).

Integration:

Policies integrating SDG goals into national frameworks ensure a holistic approach to tackling climate change and food security simultaneously.

National Policy Responses

1: Climate-Smart Agriculture (CSA) Policies:

Definition:

CSA involves practices that increase productivity, enhance resilience, and reduce greenhouse gas emissions.

Examples:

India:

The National Mission for Sustainable Agriculture promotes CSA practices, including efficient irrigation, soil health management, and agroforestry.

Brazil:

The Low-Carbon Agriculture Plan supports sustainable agricultural practices that reduce emissions and enhance resilience.

2: Agricultural Subsidies and Incentives:

Subsidies for Sustainable Practices:

Governments provide financial incentives to farmers adopting sustainable practices, such as conservation tillage, organic farming, and renewable energy use.

Examples:

European Union:

The Common Agricultural Policy (CAP) includes greening measures that reward farmers for environmentally friendly practices.

United States:

The Farm Bill offers subsidies and technical assistance for conservation practices and climate adaptation strategies.

3: Research and Development (R&D):

Investments in R&D:

Governments fund research to develop climate-resilient crop varieties, efficient irrigation technologies, and sustainable farming techniques.

Examples:

China:

Significant investments in agricultural research focus on developing drought-tolerant crops and improving water use efficiency.

Australia:

The Grains Research and Development Corporation (GRDC) supports research on climate adaptation strategies for grain production.

4: Disaster Risk Management (DRM):

Policies for Resilience:

Implementing DRM policies helps mitigate the impacts of extreme weather events on agriculture, ensuring rapid recovery and continued food production.

Examples:

Japan:

Comprehensive disaster management policies include early warning systems, infrastructure improvements, and financial support for affected farmers.

Bangladesh:

Community-based adaptation and DRM programs focus on enhancing resilience to floods and cyclones, protecting agricultural livelihoods.

Local Policy Responses

1: Community-Based Adaptation (CBA):

Definition:

CBA involves local communities in designing and implementing adaptation strategies tailored to their specific needs and conditions.

Examples:

Nepal:

Community-led initiatives in the Himalayan region focus on water management, soil conservation, and sustainable livestock practices.

Kenya:

Pastoral communities adopt traditional knowledge and modern techniques to manage grazing lands and water resources sustainably.

2: Urban Agriculture Policies:

Promotion of Urban Farming:

Local governments support urban agriculture to enhance food security, reduce the urban heat island effect, and promote green spaces.

Examples:

Cuba:

The government supports extensive urban agriculture programs, providing land, resources, and technical assistance to urban farmers.

United States:

Cities like Detroit and New York promote urban farming through grants, zoning changes, and community garden initiatives.

3: Local Food Systems and Markets:

Support for Local Producers:

Policies supporting local food systems enhance resilience by reducing dependency on distant food sources and promoting local economies.

Examples:

Italy:

The Slow Food movement encourages local food production, biodiversity preservation, and sustainable agricultural practices.

Thailand:

Local food markets and cooperatives help small-scale farmers access markets and receive fair prices for their produce.

Evaluation of Policy Effectiveness

1: Monitoring and Evaluation (M&E):

Importance of M&E:

Continuous monitoring and evaluation of policies ensure their effectiveness and provide insights for improvements.

Indicators:

Key indicators include changes in crop yields, soil health, water use efficiency, and farmers' income levels.

Examples:

Mexico:

The government uses M&E frameworks to assess the impact of CSA policies and adjust strategies based on feedback and data.

2: Stakeholder Engagement:

Inclusive Policy Development:

Engaging stakeholders, including farmers, scientists, policymakers, and civil society, in policy development ensures that policies are practical, equitable, and effective.

Examples:

Canada:

The Canadian Agricultural Partnership involves extensive consultations with stakeholders to develop and implement climate adaptation strategies.

Conclusion

Effective policy responses at international, national, and local levels are crucial for addressing the challenges posed by climate change and ensuring food security. From global agreements like the Paris Agreement to community-based adaptation initiatives, a multifaceted approach is necessary to build resilience in the agricultural sector. This chapter highlights the importance of integrated, inclusive, and adaptive policy measures, as well as the need for continuous evaluation and stakeholder engagement, to create a sustainable and secure food system in the face of climate change.

Chapter 17
Adaptation Strategies for Farmers

Introduction

As climate change continues to impact global agriculture, farmers must adopt practical strategies to adapt to the shifting conditions. This chapter provides an in-depth look at various adaptation strategies, showcasing successful measures from around the world. Through case studies and examples, we will explore how farmers are building resilience, optimizing resource use, and maintaining productivity despite the challenges posed by climate change.

Water Management Strategies

1: Efficient Irrigation Systems:

Drip Irrigation:

Delivers water directly to the plant roots, reducing evaporation and runoff.

Case Study: Israel:

Widely known for its advanced drip irrigation technology, Israel has significantly improved water use efficiency in agriculture, allowing farmers to grow crops in arid regions.

Sprinkler Systems:

Simulate natural rainfall, providing uniform water distribution.

Case Study: Nebraska, USA:

Farmers in Nebraska use center pivot irrigation systems to efficiently water large fields of corn and soybeans, optimizing water use and improving yields.

2: Rainwater Harvesting:

Definition:

Collecting and storing rainwater for agricultural use.

Case Study: India:

In regions like Rajasthan, farmers build rainwater harvesting structures such as check dams and ponds to capture and store monsoon rains for irrigation during dry periods.

3: Soil Moisture Management:

Mulching:

Applying organic or inorganic materials on the soil surface to retain moisture and reduce evaporation.

Case Study: Kenya:

Farmers in semi-arid regions use mulching to conserve soil moisture and protect crops from heat stress.

Cover Crops:

Planting cover crops to improve soil structure and moisture retention.

Case Study: Brazil:

In southern Brazil, farmers plant cover crops like legumes to improve soil health and moisture retention, enhancing the resilience of main crops.

Crop Management Strategies

1: Drought-Tolerant Crops:

Development and Adoption:

Breeding and growing crops that can withstand prolonged dry periods.

Case Study: Africa:

The introduction of drought-tolerant maize varieties through projects like the Water Efficient Maize for Africa (WEMA) has helped smallholder farmers maintain yields during droughts.

2: Crop Diversification:

Definition:

Growing a variety of crops to reduce risk and improve resilience.

Case Study: Peru:

Andean farmers cultivate a mix of traditional crops, such as potatoes, quinoa, and maize, diversifying their sources of income and food.

3: Adjusting Planting Schedules:

Timing:

Shifting planting and harvesting dates to align with changing climatic conditions.

Case Study: Bangladesh:

Farmers adjust the timing of rice planting to avoid the monsoon floods, reducing crop losses and ensuring better yields.

Soil Health Management

1: Conservation Tillage:

Definition:

Reducing or eliminating tillage to preserve soil structure and organic matter.

Case Study: USA:

In the Midwest, farmers practice no-till farming to reduce soil erosion, improve soil health, and enhance water retention.

2: Integrated Nutrient Management:

Combining Organic and Inorganic Fertilizers:

Using a mix of fertilizers to maintain soil fertility and reduce chemical input dependency.

Case Study: India:

Farmers in Punjab use integrated nutrient management practices, combining compost and synthetic fertilizers to improve soil health and crop productivity.

3: Agroforestry:

Definition:

Integrating trees and shrubs into agricultural landscapes to improve soil fertility and provide additional resources.

Case Study: Kenya:

Farmers in the highlands practice agroforestry by planting trees like Grevillea alongside crops, improving soil health and providing timber and fodder.

Livestock Management Strategies

1: Heat Stress Management:

Shade Structures and Cooling Systems:

Providing shade and cooling for livestock to reduce heat stress.

Case Study: Australia:

Farmers install shade structures and water misting systems in livestock pens to protect animals from extreme heat.

2: Improved Feeding Practices:

Supplemental Feeding:

Providing balanced diets to improve livestock health and productivity.

Case Study: Brazil:

In the Cerrado region, farmers use supplemental feeding with protein-rich forage to enhance cattle growth and milk production.

3: Breeding Resilient Livestock:

Selection of Climate-Resilient Breeds:

Breeding livestock that can thrive in changing climatic conditions.

Case Study: Africa:

Farmers in East Africa breed indigenous cattle varieties that are more resilient to drought and diseases.

Fisheries and Aquaculture Adaptation

1: Sustainable Aquaculture Practices:

Recirculating Aquaculture Systems (RAS):

Using closed-loop systems to reduce water use and improve efficiency.

Case Study: Norway:

Norwegian salmon farms use RAS to enhance sustainability and reduce environmental impacts.

2: Adaptive Fishing Practices:

Seasonal and Species Adjustments:

Adjusting fishing practices based on changing fish populations and seasons.

Case Study: Pacific Islands:

Fishermen adapt by targeting different fish species and adjusting fishing times to cope with shifting marine ecosystems.

3: Marine Protected Areas (MPAs):

Conservation Zones:

Establishing MPAs to protect critical habitats and ensure sustainable fish populations.

Case Study: Philippines:

The establishment of MPAs has helped restore fish stocks and support local fishing communities.

Financial and Social Strategies

1: Access to Credit and Insurance:

Microfinance and Crop Insurance:

Providing financial services to help farmers manage risks and invest in adaptive practices.

Case Study: India:

The Pradhan Mantri Fasal Bima Yojana (PMFBY) scheme offers crop insurance to protect farmers from weather-related losses.

2: Community-Based Adaptation:

Local Initiatives:

Engaging communities in designing and implementing adaptation strategies tailored to local conditions.

Case Study: Nepal:

Community-led water management projects in the Himalayan foothills improve irrigation efficiency and reduce conflict over water resources.

3: Farmer Education and Training:

Capacity Building:

Providing training and resources to help farmers adopt new technologies and practices.

Case Study: Uganda:

Agricultural extension services offer training on climate-smart practices, such as intercropping and soil conservation, to smallholder farmers.

Conclusion

Adaptation strategies are crucial for farmers to cope with the challenges posed by climate change. By adopting efficient water management systems, crop diversification, conservation tillage, and sustainable livestock practices, farmers can enhance their resilience and ensure sustainable food production. Case studies from around the world demonstrate the effectiveness of these strategies, highlighting the importance of local knowledge, community engagement, and continuous innovation. This chapter underscores the need for ongoing support, education, and policy frameworks to empower farmers in their adaptation efforts, ensuring food security in a changing climate.

Chapter 18
The Role of Sustainable Practices

Introduction

Sustainable agricultural practices play a crucial role in mitigating the impacts of climate change on food production. These practices not only enhance the resilience of agricultural systems but also contribute to environmental conservation and long-term food security. This chapter explores various sustainable agricultural practices, including organic farming, agroforestry, and regenerative agriculture, highlighting their benefits and providing case studies to illustrate their effectiveness.

Organic Farming

1: Definition and Principles:

Definition:

Organic farming is a method of crop and livestock production that involves the use of natural inputs and processes to enhance soil health, biodiversity, and ecological balance.

Principles:

Organic farming is based on principles such as sustainability, health, ecology, and fairness. It avoids the use of synthetic fertilizers, pesticides, and genetically modified organisms (GMOs).

2: Benefits:

Soil Health:

Organic farming practices, such as crop rotation, cover cropping, and the use of organic fertilizers (compost and manure), improve soil structure, fertility, and microbial activity.

Biodiversity:

Organic farms typically support higher levels of biodiversity, including beneficial insects, birds, and soil organisms, which contribute to pest control and pollination.

Climate Resilience:

By enhancing soil organic matter and moisture retention, organic farming increases the resilience of crops to extreme weather events, such as droughts and floods.

3: Case Study: Organic Rice Farming in the Philippines:

Background:

In the Philippines, farmers in the Ifugao region practice organic rice farming on terraced fields.

Practices:

They use traditional methods, such as the application of compost and natural pest control techniques, to maintain soil fertility and crop health.

Outcomes:

These practices have led to improved soil health, increased rice yields, and enhanced resilience to climate change.

Agroforestry

1: Definition and Principles:

Definition:

Agroforestry is an integrated approach that combines trees and shrubs with crops and/or livestock to create sustainable and productive agricultural systems.

Principles:

Agroforestry systems are designed to optimize the interactions between trees, crops, and livestock, enhancing productivity, biodiversity, and ecosystem services.

2: Benefits:

Soil and Water Conservation:

Trees and shrubs reduce soil erosion, improve water infiltration, and enhance soil fertility through leaf litter decomposition and nitrogen fixation.

Climate Mitigation:

Agroforestry systems sequester carbon in both biomass and soil, contributing to climate change mitigation.

Biodiversity and Livelihoods:

Agroforestry provides habitat for wildlife, diversifies income sources, and improves food security by producing a variety of products (fruits, nuts, timber, fodder).

3: Case Study: Agroforestry in Kenya:

Background:

In the Mount Kenya region, farmers integrate trees like Grevillea, Sesbania, and Faidherbia with crops and livestock.

Practices:

These trees provide shade, improve soil fertility, and offer additional income through timber and fruit sales.

Outcomes:

Agroforestry has improved soil health, increased crop yields, and enhanced the resilience of farming systems to climate variability.

Regenerative Agriculture

1: Definition and Principles:

Definition:

Regenerative agriculture is a holistic approach that focuses on regenerating soil health, increasing biodiversity, and improving ecosystem services to create resilient and sustainable agricultural systems.

Principles:

Core principles include minimizing soil disturbance, maintaining soil cover, promoting crop diversity, integrating livestock, and reducing synthetic inputs.

2: Benefits:

Soil Health and Carbon Sequestration:

Practices such as cover cropping, no-till farming, and compost application improve soil structure, organic matter content, and carbon sequestration.

Water Management:

Healthy soils with higher organic matter content have better water infiltration and retention, reducing the risk of drought and flooding.

Biodiversity and Ecosystem Services:

Regenerative practices support diverse plant and animal species, enhancing ecosystem services like pollination, pest control, and nutrient cycling.

3: Case Study: Regenerative Ranching in the USA:

Background:

In the Great Plains region, ranchers adopt regenerative grazing practices to improve soil health and ecosystem resilience.

Practices:

These include rotational grazing, cover cropping, and the integration of diverse plant species in pastures.

Outcomes:

Regenerative ranching has led to improved soil health, increased forage production, and enhanced resilience to drought.

Integrating Sustainable Practices

1: Combination of Practices:

Synergies:

Combining different sustainable practices can create synergies that enhance overall system resilience and productivity. For example, integrating agroforestry with organic farming can improve soil health, biodiversity, and crop yields simultaneously.

2: Local Adaptation:

Context-Specific Approaches:

Sustainable practices must be adapted to local conditions, considering factors such as climate, soil types, and socio-economic contexts. This ensures that the practices are effective and feasible for local farmers.

3: Policy Support and Incentives:

Government and Institutional Support:

Policies and incentives that promote sustainable practices, such as subsidies for organic farming, payments for ecosystem services, and technical assistance, are crucial for widespread adoption.

Examples:

European Union:

The Common Agricultural Policy (CAP) supports sustainable farming practices through greening measures and agri-environmental schemes.

Costa Rica:

The government provides payments for ecosystem services to farmers who adopt sustainable practices, such as agroforestry and reforestation.

Challenges and Future Directions

1: Challenges:

Knowledge and Training:

Lack of knowledge and training on sustainable practices can hinder adoption. Extension services and farmer education programs are essential.

Initial Costs:

The initial investment required for some sustainable practices, such as setting up irrigation systems or purchasing cover crop seeds, can be a barrier for smallholder farmers.

Market Access:

Access to markets for sustainably produced goods, such as organic products, can be limited. Developing market infrastructure and certification systems can help.

2: Future Directions:

Research and Innovation:

Continued research and innovation in sustainable agriculture, including the development of climate-resilient crop varieties and advanced soil management techniques, are crucial.

Scaling Up:

Efforts to scale up sustainable practices through policy support, public-private partnerships, and community-based initiatives will enhance their impact.

Global Collaboration:

International collaboration and knowledge exchange on sustainable practices can accelerate their adoption and adaptation to different contexts.

Conclusion

Sustainable agricultural practices, such as organic farming, agroforestry, and regenerative agriculture, are vital for mitigating the impacts of climate change and ensuring long-term food security. These practices enhance soil health, increase biodiversity, sequester carbon, and improve resilience to climate variability. By integrating these practices and supporting their adoption through policies and incentives, we can create more resilient and sustainable agricultural systems. This chapter highlights the importance of sustainable practices in building a climate-resilient future for agriculture and underscores the need for continued innovation, education, and policy support.

Chapter 19
Future Scenarios and Predictions

Introduction

Understanding the potential future impacts of climate change on global food production is critical for developing effective strategies to mitigate risks and ensure food security. This chapter explores various future scenarios and predictions, examining the likely outcomes based on current trends, scientific models, and potential mitigation and adaptation strategies. By analyzing these scenarios, we aim to provide a comprehensive view of the challenges and opportunities that lie ahead in the quest to sustain global food production in a changing climate.

Future Scenarios Based on Current Trends

1: Business-as-Usual Scenario:

Continued Emissions Growth:

If greenhouse gas emissions continue to rise at the current rate, global temperatures could increase by 4-5°C by the end of the century.

Impact on Crop Yields:

High temperatures, more frequent extreme weather events, and changing precipitation patterns will significantly reduce crop yields. Staple crops such as wheat, maize, and rice are particularly vulnerable.

Water Scarcity:

Increased temperatures and altered precipitation patterns will exacerbate water scarcity, impacting irrigation and rain-fed agriculture, especially in already water-stressed regions.

2: Moderate Emission Reduction Scenario:

Stabilized Emissions:

With moderate efforts to reduce emissions, global temperatures may rise by approximately 2-3°C by 2100.

Partial Crop Resilience:

Some crops may benefit from elevated CO_2 levels, but the overall negative impacts of climate change, such as heat stress and water scarcity, will outweigh these benefits.

Regional Variability:

The impact on agriculture will vary significantly by region. Some areas may experience slight improvements in growing conditions, while others will face severe challenges.

2: Aggressive Mitigation Scenario:

Significant Emission Reductions:

With aggressive mitigation efforts, including transitioning to renewable energy and implementing carbon capture technologies, global temperature rise could be limited to 1.5°C.

Increased Resilience:

Reduced climate change impacts will improve the resilience of agricultural systems. Adaptation strategies, such as breeding climate-resilient crops and improving water management, will be more effective.

Sustainable Practices:

The widespread adoption of sustainable agricultural practices, such as regenerative agriculture and agroforestry, will enhance soil health, biodiversity, and crop yields.

Potential Outcomes for Major Agricultural Regions

1: Africa:

Business-as-Usual:

Sub-Saharan Africa will face severe food insecurity due to reduced crop yields, increased droughts, and water scarcity. Staple crops like maize and sorghum will be particularly affected.

Moderate Mitigation:

Adaptation strategies, such as drought-resistant crops and improved irrigation, will help mitigate some impacts. However, food insecurity will remain a significant challenge.

Aggressive Mitigation:

With significant mitigation efforts and adaptation measures, food security can improve. Sustainable practices and enhanced infrastructure will support resilience.

2: Asia:

Business-as-Usual:

South and Southeast Asia will experience reduced rice and wheat yields due to rising temperatures, erratic monsoon patterns, and increased flooding. Coastal regions will face salinity intrusion.

Moderate Mitigation:

Improved water management and crop diversification will help manage some impacts. However, the region will still face significant challenges in maintaining food security.

Aggressive Mitigation:

Enhanced climate resilience through sustainable practices, improved infrastructure, and robust adaptation strategies will support food security in the region.

3: Europe:

Business-as-Usual:

Southern Europe will face severe droughts and heatwaves, reducing crop yields and water availability. Northern Europe may experience some initial benefits from longer growing seasons.

Moderate Mitigation:

Adaptation strategies, such as improved irrigation and crop selection, will mitigate some impacts. However, southern Europe will still face significant challenges.

Aggressive Mitigation:

Increased resilience through sustainable practices, technological innovation, and effective policy measures will support food production across Europe.

4: Americas:

Business-as-Usual:

North America will face increased heat stress on crops, reduced water availability, and more frequent extreme weather events. Latin America will experience reduced crop yields, increased droughts, and deforestation impacts.

Moderate Mitigation:

Improved water management, climate-resilient crops, and sustainable land use practices will help mitigate some impacts. However, significant challenges will remain.

Aggressive Mitigation:

Enhanced resilience through sustainable practices, technological innovation, and effective policy measures will support food production across the Americas.

Mitigation and Adaptation Strategies

1: Mitigation Strategies:

Renewable Energy Transition:

Shifting to renewable energy sources, such as solar and wind, to reduce greenhouse gas emissions and slow climate change.

Carbon Capture and Storage:

Implementing carbon capture technologies to remove CO_2 from the atmosphere and store it underground.

Sustainable Agriculture:

Promoting sustainable agricultural practices, such as organic farming, agroforestry, and regenerative agriculture, to sequester carbon and reduce emissions.

2: Adaptation Strategies:

Climate-Resilient Crops:

Breeding and adopting crops that are resistant to heat, drought, and pests to maintain yields under changing conditions.

Water Management:

Implementing efficient irrigation systems, rainwater harvesting, and integrated water management practices to address water scarcity.

Soil Health Improvement:

Enhancing soil health through practices such as cover cropping, reduced tillage, and organic amendments to improve resilience and productivity.

Technological Innovations

1: Precision Agriculture:

Definition: Using technology, such as GPS, sensors, and drones, to optimize farming practices and resource use.

Benefits:

Precision agriculture improves efficiency, reduces input costs, and enhances crop yields. It also supports data-driven decision-making for better adaptation to climate variability.

2: Genetic Engineering:

Definition:

Developing genetically modified crops that are more resilient to climate change impacts, such as drought, heat, and pests.

Benefits:

Genetic engineering can enhance crop resilience, improve yields, and reduce reliance on chemical inputs, supporting sustainable food production.

3: Climate Modeling and Forecasting:

Definition:

Using advanced climate models to predict future climate scenarios and their impacts on agriculture.

Benefits:

Improved forecasting supports better planning and decision-making, helping farmers and policymakers develop effective adaptation and mitigation strategies.

Policy and Governance

1: International Agreements:

Paris Agreement:

A global framework to limit global warming to well below 2°C and pursue efforts to limit it to 1.5°C. Countries commit to reducing emissions and enhancing resilience through national climate action plans (NDCs).

2: National Policies:

Climate-Smart Agriculture:

Governments implement policies to promote climate-smart agriculture, which integrates climate change adaptation and mitigation into agricultural development.

Subsidies and Incentives:

Providing financial incentives for sustainable practices, such as subsidies for organic farming, payments for ecosystem services, and support for technological innovation.

3: Local and Community Initiatives:

Community-Based Adaptation:

Engaging local communities in designing and implementing adaptation strategies tailored to local conditions.

Public-Private Partnerships:

Collaborating with private sector stakeholders to invest in sustainable agriculture, infrastructure, and innovation.

Conclusion

Future scenarios and predictions highlight the significant challenges and opportunities that climate change presents to global food production. By analyzing different scenarios based on current trends and potential mitigation and adaptation strategies, we can better understand the potential outcomes and prepare for the future. Aggressive mitigation efforts and the widespread adoption of sustainable practices are essential to enhance resilience, ensure food security, and mitigate the impacts of climate change on agriculture. This chapter underscores the importance of continued research, innovation, policy support, and global collaboration in building a sustainable and resilient food system for future generations.

Chapter 20
Conclusion
Towards a Resilient Food System

Introduction

As we conclude our exploration of the intricate relationship between climate change and global food production, it is evident that building a resilient food system is paramount for ensuring food security in the face of a rapidly changing climate. This chapter summarizes the key points discussed throughout the book and outlines a vision for a resilient global food system. Emphasizing the importance of coordinated action and long-term planning, we aim to inspire a collective commitment to safeguarding food security for future generations.

Summary of Key Points

1: Climate-Food Nexus:

Interconnectedness:

The delicate balance between climate and food production has been a cornerstone of human civilization. Understanding this nexus is crucial for addressing the impacts of climate change on agriculture.

2: Historical Perspectives:

Lessons from the Past:

Historical climate variations have profoundly impacted agricultural practices. Ancient civilizations adapted to changing climates, offering valuable lessons for current and future challenges.

3: Current State of Global Food Production:

Critical Regions and Crops:

The current state of global food production is sustained by major crops and livestock, with specific regions playing vital roles in global food security.

4: Science of Climate Change:

Foundational Understanding:

The greenhouse effect, carbon cycle, and the role of human activities are fundamental to understanding the science behind climate change and its impacts on agriculture.

5: Impacts on Crop Yields and Livestock:

Temperature and Extreme Weather:

Rising temperatures and extreme weather events adversely affect crop yields and livestock, posing significant challenges to food production.

6: Water Scarcity and Soil Health:

Critical Issues:

Water scarcity and declining soil health are major challenges exacerbated by climate change. Sustainable water management and soil conservation are essential for resilience.

7: Pests and Diseases:

Increased Threats:

Changing climate conditions alter the prevalence and distribution of agricultural pests and diseases, necessitating effective crop protection strategies.

8: Regional Case Studies:

Global Impact:

Climate change affects food production differently across regions. Case studies from Africa, Asia, Europe, and the Americas highlight unique challenges and adaptation strategies.

9: Technological Innovations:

Role of Technology:

Innovations such as precision agriculture, genetically modified crops, and sustainable farming practices are pivotal in addressing climate challenges.

10: Policy Responses:

Government and Institutional Action:

Policy responses at international, national, and local levels are critical for addressing food security in the face of climate change.

11: Adaptation and Sustainable Practices:

Resilience Building:

Practical strategies for farmers, sustainable agricultural practices, and the integration of diverse approaches are key to building resilience.

12: Future Scenarios and Predictions:

Potential Outcomes:

Various future scenarios underscore the importance of aggressive mitigation efforts and adaptation strategies to ensure a sustainable future for global food production.

Vision for a Resilient Food System

1: Holistic Approach:

Integrated Strategies:

Building a resilient food system requires a holistic approach that integrates sustainable agricultural practices, technological innovation, and effective policy measures.

2: Sustainable Agriculture:

Promoting Practices:

Widespread adoption of sustainable practices, such as organic farming, agroforestry, and regenerative agriculture, will enhance soil health, biodiversity, and climate resilience.

3: Technological Innovation:

Advancing Technology:

Continued research and development in agricultural technologies, including precision farming, genetic engineering, and climate modeling, are essential for adaptation and mitigation.

4: Policy and Governance:

Supportive Frameworks:

Governments and institutions must create supportive policies and incentives to promote sustainable practices and invest in resilient infrastructure.

5: Community and Global Collaboration:

Collective Action:

Collaboration at local, national, and international levels is crucial. Engaging communities, fostering public-private partnerships, and facilitating global knowledge exchange will drive progress.

6: Education and Awareness:

Empowering Stakeholders:

Education and awareness programs for farmers, policymakers, and the general public are vital for promoting sustainable practices and fostering a culture of resilience.

7: Equity and Inclusion:

Ensuring Access:

Ensuring equitable access to resources, technology, and markets for smallholder farmers and vulnerable communities is essential for building a resilient food system.

8: Long-Term Planning:

Visionary Leadership:

Long-term planning and visionary leadership are necessary to address the complex challenges posed by climate change. This includes setting ambitious goals, monitoring progress, and adjusting strategies as needed.

Call to Action

1: Urgency and Commitment:

Immediate Action:

The urgency of addressing climate change impacts on agriculture cannot be overstated. Immediate and sustained action is required to mitigate risks and build resilience.

2: Global Responsibility:

Shared Responsibility:

Ensuring global food security in the face of climate change is a shared responsibility. Governments, institutions, businesses, and individuals must work together towards this common goal.

3: Innovative Solutions:

Embracing Innovation:

Embracing innovative solutions and leveraging technology will be key to overcoming the challenges posed by climate change and transforming the global food system.

4: Sustainable Future:

Vision for the Future:

Our vision for a resilient food system is one where sustainable practices, technological advancements, supportive policies, and global collaboration converge to ensure food security for future generations.

Conclusion

As we navigate the complexities of climate change and its impacts on global food production, it is clear that a resilient food system is essential for sustaining life on Earth. By embracing sustainable practices, advancing technological innovation, enacting supportive policies, and fostering global collaboration, we can build a resilient and equitable food system that ensures food security for future generations.

This book has highlighted the challenges and opportunities that lie ahead, underscoring the importance of coordinated action and long-term planning. Together, we can create a sustainable and resilient future for agriculture and humanity.

www.ingramcontent.com/pod-product-compliance
Lightning Source LLC
Chambersburg PA
CBHW071925210526
45479CB00002B/560